T0275562

Autonomic Network Management Principles

Form Concepts to Applications

Autonomic Network Management Principles

Form Concepts to Applications

Nazim Agoulmine

AMSTERDAM • BOSTON • HEIDELBERG • LONDON
NEW YORK • OXFORD • PARIS • SAN DIEGO
SAN FRANCISCO • SINGAPORE • SYDNEY • TOKYO

Academic Press is an imprint of Elsevier

Academic Press is an imprint of Elsevier
30 Corporate Drive, Suite 400, Burlington, MA 01803, USA
525 B Street, Suite 1900, San Diego, California 92101-4495, USA
84 Theobald's Road, London WC1X 8RR, UK

Library of Congress Cataloging-in-Publication Data
Agoulmine, Nazim.
Autonomic network management principles / Nazim Agoulmine.
    p. cm.
  ISBN 978-0-12-810199-5
1. Computer networks–Management. I. Title.
  TK5105.5.A36 2010
  004.6–dc22

                                                2010033247

British Library Cataloguing-in-Publication Data
A catalogue record for this book is available from the British Library.

For information on all Academic Press publications
visit our website at *www.elsevierdirect.com*

Printed in the United States of America
11 12 13   10 9 8 7 6 5 4 3 2 1

Contents

The development of network technologies and networked services has been tremendous the last decade and is expected to be more intensive in the near future with the technological progress in the field of radio communications, data communication, protocols, software engineering, applications and so on. In this plethora of technologies, operators as well as administrators are facing more and more complex failures and dysfunctions of their networks. The management of these problems are reaching the limit of human experts requiring skills and knowledge that are expensive and difficult to acquire. Several directions are under investigation to find solution to these problems without human intervention. Among these directions, Autonomic Network Management aims at introducing more self-management in existing equipments, protocols and services to allow networks to self-recover and to autonomically adjust to new situation. Autonomic Network Management takes its essences from several advances in researches from networking, computing, intelligent systems, data engineering, etc. It requires essential cross disciplinary research to overcome the challenges posed by existing and future complex and heterogeneous communication, networking and applications technologies.

This book aims at presenting the basic principle of autonomics and how they could apply to build Autonomic Network Management Architecture through several experiences presented by several well recognized experts in the area that have contributed to significant advances in the area. These contributors from research laboratory and industry present their results in different application areas to provide the reader with up-to-date information about this novel approach in network management. The contributions aims to cover wide range of management domains as well as reference architectures inspired from autonomic computing, biology, and intelligent decision systems. Hopefully, the contributions covers all the interests of specialists, researchers and students in the area tackling different levels of network architecture from the core network management to wire and wireless access networks management including promising technologies such as sensor networks.

Besides the treatment of the theoretical foundations of autonomics, this book puts special emphasis on the presentation of algorithms and realistic solutions, which are fundamental to help engineers and researcher to design successful practical applications to solve real management. The contributions show which adjustments to the respective applications area are necessary and how it is possible to implement them to improve the manageability of the network and supported services. This book also discusses emerging trends in Autonomic Network Management that could form the basis of the design of Next Generation Networks.

We hope this book will be very useful for all researchers and practitioners from the field of network and service management as well as graduate students, who want to devote themselves to the design of future innovative IT management, autonomic computing, autonomic networking, resources management, sensor network management, or related problems and want to gain a deeper understanding of the application of autonomic concepts and techniques in the area of networking.

About the Authors

Nazim Agoulmine, Ing, Msc, Ph.D. is a Full Professor at the University of Evry, France and a WCU distinguished Visiting Professor at Postech, Korea. He is leading a research group in the IBISC Lab. His research interests include autonomic network and service management architectures, optimization techniques for self-management in wired, wireless and sensors networks, utility theory applied to autonomic behaviors, and biologically inspired models for self-organization of communication networks.

Elyes Lehtihet is a postdoctoral researcher with the IBISC Laboratory in France. He received his PhD from the University of Evry in 2009, his Master's degree from the University of Marne La Vallée in 2004, and his Engineering diploma from the University of Annaba (2003). He is working on the reuse of semantic Web technologies for autonomic network and service management.

Hajer Derbel holds a Ph.D. in computer science from the University of Evry, Val d'Essonne, France. She received her Engineering diploma in 2003 from ENIS (Tunisia) and her Master's degree in 2004 from UTT, France. Her research interests are in applying Utility Theory in network management and control to build future User Aware autonomic networks.

Ibrahim Al-Oqily holds a Ph.D. degree in computer science from the University of Ottawa, Canada. He is currently an assistant professor at the Faculty of Prince Al-Hussein Bin Abdallah II for Information Technology (IT), Hashemite University. His current research interests include overlay networks management, mobile computing, autonomic networks, service composition, and policy-based networks management.

Ahmed Karmouch received his Ph.D. degree in computer science from the University of Paul Sabatier, Toulouse, France, in 1979. He is a Professor at the School of Information Technology and Engineering, University of Ottawa. His current research interests are in distributed multimedia systems and communications, mobile computing, autonomic networks, context aware communications, ambient networks and WiMax based applications and services.

Prométhée Spathis is an Associate Professor at the University of Pierre et Marie Curie (Paris 6 University) in France. His research interests as a member of the Nework and Performance Analysis group are autonomous networks and content centric networking. Dr. Spathis has been involved in several European projects on autonomics such as FP6 IP ANA project, the ACCA coordination action, and several other national research initiatives.

Marco Bicudo joined the Network and Performance Analysis team at the University of Pierre et Marie Curie (Paris 6 University) in 2008 where he is working as a research engineer in the frame of FP6 projects. He received a M.Sc. degree in electrical and computer engineering from Universidade Federal do Rio de Janeiro in 2006. He is the Engineer Chief of Operations for PlanetLab Europe.

Brendan Jennings, BEng, PhD, is a Senior Research Fellow with the Telecommunications Software & Systems Group at Waterford Institute of Technology, Ireland. His research interests include autonomic network management, charging and billing, and performance management.

Kevin Chekov Feeney, BA (Mod.), PhD, is a Research Fellow with the Knowledge and Data Engineering Group (KDEG) in the School of Computer Science and Statistics at Trinity College, Dublin, Ireland. He has published widely on the topic of policy-based network management and has previously held positions as a software developer, designer, and architect.

Rob Brennan, BSc, MSc, PhD, is a Research Fellow with the Knowledge and Data Engineering Group (KDEG) in the School of Computer Science and Statistics at Trinity College, Dublin, Ireland. His research interests include semantic mapping, distributed systems, and the application of semantics to network and service management.

Sasitharan Balasubramaniam, BEng, MEngSc, PhD, is a Research Fellow with the Telecommunications Software & Systems Group at Waterford Institute of Technology, Ireland. His research interests include bio-inspired autonomic network management, as well as sensor and ad hoc networking.

Dmitri Botvich, BSc, PhD, is a Principal Investigator with the Telecommunications Software & Systems Group at Waterford Institute of Technology, Ireland. His research interests include bio-inspired autonomic network management, security, trust management, sensor, and ad-hoc networking, queuing theory, and mathematical physics.

Sven van der Meer (vdmeer@ieee.org) is a Senior Research Fellow with the Telecommunications Software & Systems Group at Waterford Institute of Technology, Ireland. He leads Irish and European programs in network management developing strong industrial links, serves in international conferences, and is active in standardization.

Masayuki Murata is a Professor at the Graduate School of Information Science and Technology, Osaka University, Japan. His research interests include computer communication network architecture.

Naoki Wakamiya is an Associate Professor at the Graduate School of Information Science and Technology, Osaka University, Japan. His interests include self-organization of overlay networks, sensor networks, and mobile ad-hoc networks.

Dr. Kenji Leibnitz is a senior researcher at the National Institute of Information and Communications Technology, Japan. His interests are in biologically inspired models for self-organization of communication networks.

Dr. Hiroshi Wada is a researcher at National ICT, Australia, and a conjoint lecturer at the School of Computer Science and Engineering, University of New South Wales. His research interests include model-driven software development, performance engineering, and search-based software engineering.

Pruet Boonma received a Ph.D. in computer science from the University of Massachusetts, Boston, in 2010. He is currently a lecturer in the computer engineering department at Chiang Mai University, Thailand. His research interests include autonomous adaptive distributed systems and metaheuristic optimization algorithms.

Junichi Suzuki received a Ph.D. in computer science from Keio University, Japan, in 2001. He is currently an Associate Professor of computer science at the University of Massachusetts, Boston. His research interests include autonomous adaptive distributed systems, wireless sensor networks, and model-driven software/performance engineering.

Ehab Al-Shaer is the Director of the Cyber Defense and Network Assurability (CyberDNA) Center in the School of Computing and Informatics at the University of North Carolina at Charlotte. His primary research areas are network security, security management, fault diagnosis, and network assurability.

Latifur R. Khan is currently an Associate Professor and Director of the state-of-the-art DBL@UTD, UTD Data Mining/Database Laboratory in the Computer Science Department at the University of Texas at Dallas (UTD). Dr. Khan's research areas cover data mining, multimedia information management, semantic Web, and database systems.

Mohammad Salim Ahmed is currently pursuing his Ph.D. in computer science under Dr. Latifur R. Khan at the University of Texas at Dallas (UTD). His area of research includes network security and text mining.

John Strassner is a Professor of computer science and engineering at the Pohang University of Science and Technology (POSTECH), and leads the Autonomic Computing group in POSTECH's World Class University, IT Convergence Engineering Division. Previously, he was a Motorola Fellow and Vice President of Autonomic Research. Before that, he was the Chief Strategy Officer for Intelliden and a former Cisco Fellow. John is the Chairman of the Autonomic Communications Forum and the past chair of the TMF's NGOSS SID, metamodel, and policy working groups. He has authored two books, written chapters for five other books, and has been co-editor of five journals dedicated to network and service management and autonomics. John is the recipient of the IEEE Daniel A. Stokesbury award for excellence in network management and the Albert Einstein award for his achievements in autonomics and policy management. He is a TMF Distinguished Fellow and has authored over 240 refereed journal papers and publications.

Acknowledgement

This project has been planned quite a while ago, and after extensive planning, invitations, and revisions, the book has finally taken shape. Similar to all technology that surrounds us, there are uncertainties with new technologies. Such uncertainties usually arises from questions, such as how such technologies will be accepted and will it create a sustainable impact in the area, or is it just a fashionable hype that creates a short period of high pitched buzz. However, the concept of *Autonomic Networks* has taken a positive uptake, and continues to do so. The main driving force behind this is the big impact autonomic has integrated into systems that support our everyday lives. Therefore, as the complexity of systems surrounding us is amplifying, we are seeing a higher reliance towards autonomics, and this is being seen as a strong potential for communication networks of the future.

Writing and organizing this book has proven to be a very difficult and challenging exercise, and this is due to the selection process of high quality contributions from renowned contributors. However, this task has simplified tremendously, thanks to the support from numerous well known specialists in the area who have contributed to this book. Therefore, I would like to thank all the authors who have contributed to this book, without whom it would have been impossible to achieve. I would also like to pass my special thanks to Dr. Sasitharan Balasubramaniam who particularly helped me in the various stages of the book.

I would also like to pass my special thanks to my editor Tim Pitts for his patience since the beginning of this project, and having confidence in my proposal of this book. Thanks also to Melanie Benson for all her support, and thanks as well to Naomi Robertson as well as to Kirubhagaran Palani for their huge help in editing this book.

Lastly, I would like to thank my parents, who have always provided encouragement to me in all projects that I have embarked on over and over years.

In addition, the final version of this manuscript was prepared while Dr. N. Agoulmine was a WCU Distinguished Visiting Professor at POSTECH, Pohang, South Korea, under the auspices of the Ministry of Education, Science, and Technology Program—Project no. R31-2008-000-10100-0—and their support is gratefully acknowledged.

Chapter 1

Introduction to Autonomic Concepts Applied to Future Self-Managed Networks

Nazim Agoulmine

DEFINITION AND SCOPE

Network area has seen tremendous changes during the last decade through several waves. The first wave has been the Internet, which has completely changed the way network services are provided and helped global usage networked applications worldwide. The second wave came with the cellular and wireless technologies that have allowed the provisioning of telephony and data services anyplace, anytime. Cellular technologies have also drastically changed our behavior, allowing any person to be able to make phone calls from anywhere. They have also helped poor countries to develop efficient and cost-effective telephony infrastructure quickly, which was not possible with wire technologies. The third wave has surprisingly come from the POTS (Plain Old Telephony System) last mile access. Indeed, development of DSL (Digital Subscriber Line) technologies has enabled operators to provide high-speed IP access through a telephone line without having to pay for optical fiber installation and by taking advantage of existing telephone lines. This has allowed the emergence of the so-called Triple Play Services to the home (IP data access, Telephony of IP, and TV over IP). The fourth ware is probably under way with the convergence of the services Triple Play plus Mobile (sometimes called Quadruple Play) but also All-in-One emerging services such as P2P, social networking, and presence.

In this ideal picture of technology development, there are many behind-the-scenes issues. Indeed, slowly but steadily, the users' focus has changed from the high-speed network to value-added services. Users no longer care about the technologies of networks that are deployed (as they are lost in the different terminologies) but rather are more concerned about the quality of services they can use anywhere, anytime and at an acceptable cost. Internal architectures and protocol are only of interest to experts and are not important to customers whose

Autonomic Network Management Principles. DOI: 10.1016/B978-0-12-382190-4-00001-2

sole concern is the benefit they can get from the services. For the operator, however, internal architecture, protocols and so on are very important as they drive their capabilities to respond to customers' needs. Like any other system but with hundreds of orders of magnitude of complexity, the network has to change and evolve regularly. Operators are constantly integrating new services, new components, and new technologies without interrupting the ongoing one to fulfill new customers' needs, resolve problems, increase capacity, and the like. Every year operators' networks have become larger and more complex, dealing with a numerous heterogeneous sources (hardware, software, services, etc.). Relations between operators to allow services to span their networks to fulfill the end-to-end businesses of their customers have also added to this complex picture. And the picture becomes even more complicated as operators and equipment builders evolve in a highly revenue-generative but also deregulated area where they are all pushed to reduce their cost while facing newly entering competitors. These competitors are more aggressive as they usually enter with the latest technology without having to support several years of investments in older technologies.

On one side, the network complexity is increasing the need for more expertise and for more efforts from highly skilled people to maintain the infrastructure (Figure 1.1), and on the other side, the deregulated market is pushing for more competition and lower prices. It seems that the old operators and constructors have somehow found their way in the competition, though at a higher cost than the newly entered actors. The cost is of course related to CAPEX (Capital Expenditure) in the acquisition of new equipments and also OPEX (Operational

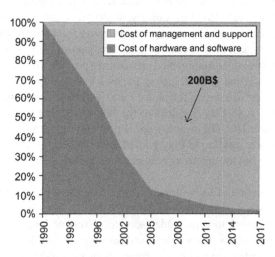

FIGURE 1.1 Evolution of the Costs of Hardware and Software and Their Management. The cost of management and support of computing and networking system has increased drastically in the past year due to the complexity of the technologies requiring ever more skilled engineers and administrators.

Cost) to control and maintain this increasingly sophisticated equipment as well as existing infrastructure and corresponding line of products.

This situation is very difficult for old operators and constructors as they could rely on huge benefits in the past; however, if a similar situation arises in the future, it will be much harder for them to survive as they will not have the necessary resources to compete.

We can easily imagine the emergence of a new network technology in the future that will be cheaper to deploy and maintain than the existing ones and could therefore supersede all existing technologies. If new actors enter the market with this technology, they will be able to provide better services at a lower cost while the remaining operators will still need to amortize their costly existing equipment and associated human resources.

Operators and constructors recognized this problem several years ago not only in the IT world but also in the networking area. Unfortunately, network and system management solutions made significant advances and are no more capable to deal with the increasing complexity; they still rely on very expensive and rare human experts to solve problems, which themselves are beyond the capacities of the experts. Many problems also arise from these experts' intervention, such as misconfigurations (wrong configuration, tuning, etc.). These misconfigurations are among the most complex problems to solve; they are very difficult both to understand and locate and therefore to fix. Operators now understand that it is vital for them to control this increased, uncontrollable operational cost (OPEX) (including the deployment cost) deploying breaking approaches.

The only response to this unsustainable situation is innovation in the way networks are managed and controlled. It is necessary to develop new networks that are able to automatically adapt their configurations to the increases and changing requirements of end users and service providers. Soon, we'll see drastic developments in the end users' services with the introduction of high-speed access networks that are either fixed with the deployment of FFTH or wireless with LTE and WiMAX technologies. Future networks need to be more flexible, capable of reorganizing in an autonomic way when new types of equipment or services are introduced, reducing the need for human intervention and consequently associated costs. Future networks should be able to improve their performances when needed to respond to unusual changes in the traffic pattern. The innovation should help to design new types of equipments, protocols, and network architectures and even services that are able to be self-managed, to reduce the operational burden on the operators by themselves making decisions in terms of configuration, optimization, and the like.

If networks and services are able to exhibit some level of autonomy that will allow them to themselves solve their problems in any context, then the operator will be able to reduce the need for intervention by human experts and therefore reduce their operational costs (OPEX). It is time that significant progress be made in how to manage and control these complex infrastructures at the early stage of their design.

Many initiatives have been launched to push toward innovations in this area. These initiatives have different names, but all converge to the emergence of a new generation of intelligent equipments, networks, and services that are able to exhibit self-properties. These initiatives are variously named—for example, Autonomic Communication (AC), Autonomic Networks (AN), Autonomic Network Management (ANM), Self-Managed Networks (SFN), Situated Networks (SN). Differences in the focus of the various approaches can explain roughly the differences in the terminology, but all of them have one thing in common: They all seek to introduce self-adaptive capabilities in the network, avoiding human interventions as much as possible.

EPIDEMIOLOGICAL DEFINITION OF AUTONOMICS

According to the Oxford English Dictionary, "autonomic" is the adjective derived from "autonomy," meaning self-governing or independent [1]. With respect to physiology, the autonomic nervous system is that part of the human body that functions independently of the will. The Cambridge physiologist John Newport Langley (1852–1925) was the first scientist to apply this term, in his publication in the *Journal of Physiology* [2] in 1898: "I propose the term 'autonomic nervous system' for the sympathetic system and allied nervous system of the cranial and sacral nerves, and for the local nervous system of the gut."

The autonomic nervous system (ANS) has of course an important role in the biological system as it regulates involuntary activity in the body by transmitting motor impulses to cardiac muscle, smooth muscle, and the glands. The ANS controls all the vital muscular activities of the heart and of the circulatory, digestive, respiratory, and urogenital systems. The autonomic nervous system governs our heart and body temperature, thus freeing our conscious brain to deal with higher level activities.

THE NEED FOR AUTONOMIC SYSTEMS

The tremendous complexity of computing systems during the last decades has exponentially increased the management and operation expenses of these systems. Operators and system administrators are envisioning IT systems that can self-govern and solve their configuration problems to achieve objectives in an autonomic. Although this idea has been the subject of many works in the area of artificial intelligence (AI), what is different today is that on one hand technologies have evolved in an impressive way, allowing new types of solutions, and on the other hand the operator's requirements are much more precise than the general case AI tried to solve in the past.

IBM was the first company to use the business keyword autonomic computing (AC), aiming at developing a new generation of intelligent computer systems [6]. Paul Horn of IBM introduced AC in October 2001 in a pioneer work in the new wave of autonomics. AC is used to embody self-managing. In the IT industry, the term *autonomic* refers to the ability of a component to self-manage based on internal stimuli. More precisely, "autonomic" means the act of acting and occurring involuntarily—that is, systems that are able to manage themselves based on high-level administration objectives. With AC human administrators would no longer need to deal with low-level management and could then concentrate on the higher level management process.

AUTOMATIC, AUTONOMOUS, AND AUTONOMIC SYSTEMS

There are some semantic differences between autonomic, autonomous, and automatic that can be summarized in the following definitions [see Collins, *Educational Dictionary Millennium Edition,* 2000] [21]

- **Definition of Automatic (Adv Automatically):** An autonomic action is an action that is performed from force of habit or without any conscious thought. Many examples can of course be found in the bio-system or artificial system. The human body is able to perform a number of reflex or involuntary actions, while an automatic system is designed to perform some specific actions as a consequence of some occurring events or known problems. Automatic systems do not have any knowledge outside the predefined one and no ability to extend it. An automatic system will always exhibit the same behaviors for the same input. In automatic theory, this behavior is called transfer function. Even though the word "automatic" comes from the Greek word *automatous,* which means acting for one's own will, it has become associated with mechanical terms that are predefined and not running of one's free will [22].
- **Definition of Autonomous (Adv Autonomously):** An autonomous system is a system that exhibits a large degree of self-governance. This system takes its decision without referring to any external entities and in complete independence. The autonomous entity defines its own rules and principles to achieve its own goals. An autonomous behavior is the ultimate freedom.
- **Definition of Autonomic (Adv Autonomically):** The word "autonomic" suggests the idea of self-governance within an entity based on internal policies and principles, which can also be described as autonomics. Autonomic relating to the autonomic nervous system (ANS) is based on internal stimuli that trigger involuntary responses. In medical terms, autonomic means self-controlling or functionality independent.

The autonomic system in the IT world uses a holistic approach to the design of highly distributed and complex distributed computing environments resulting in self-managed systems. Autonomic systems are inspired by ANS, where ANS manages the important bodily functions devoid of any conscious involvement. It describes a system in which humans define goals as input to a self-managing autonomous distributed and more likely heterogeneous system.

IBM'S APPLICATION OF AUTONOMICS TO COMPUTERS

IBM has introduced an evolutionary process to evaluate the level of autonomic behavior in a computing system. The approach also defines the basic self-managing concept. This evolutionary process defines different levels of evolution of the system management process and describes how this process evolves with the adoption of autonomic concepts and technologies [5]. The five levels of evolution toward effecting a fully autonomic system are as follows:

- *Level 1 or Basic Level.* In this system, the operator needs to monitor and configure each element manually during its entire life cycle from installation to uninstallation.
- *Level 2 or Managed Level.* In this level, the system operator can take advantage of a set of system management technologies to monitor multiple systems and system elements simultaneously using management consoles with appropriate human machine interfaces.
- *Level 3 or Proactive Level.* Advances in analytical studies of the system allow the development of a system with predictive capacities that allow analyzing gathered information and identifying and predicting problems and therefore propose appropriate solutions to the operator of the system for deployment.
- *Level 4 or Adaptive Level.* In this level, the system is not only able to gather monitored information and predict situations but also to react automatically in many situations without any human intervention. This is based on a better understanding of system behavior and control. Once knowledge of what to perform in which situation is specified, the system can carry out numerous lower level decisions and actions.
- *Level 5 Autonomic Level.* this is the ultimate level where the interactions between the humans and the systems are only based on high-level goals. Human operators only specify business policies and objectives to govern systems, while the system interprets these high-level policies and responds accordingly. At this level, human operators will trust the system in managing themselves and will concentrate solely on higher level business.

These levels range from a totally operator-managed system to an autonomically managed system based on high-level objectives. The goal of the autonomic research domain is to achieve a fully Level 5 system by researching methods and innovative approaches leading to the development of an autonomic system.

IBM AUTONOMICS COMPUTING

IBM breaks down autonomic self-management into four categories: self-configuration, self-healing, self-optimization, and self-protection. The overall goal of these concepts is to deliver a self-managing solution that is proactive, robust, adaptable, and easy to use. In accordance with Roy Sterritt's Autonomic Computing Tree 0[7], the Autonomic Computing initiative will require four primary objectives known as CHOP [8]: Self-Configuration, Self-Healing, Self-Optimization, and Self-Protecting. Figure 1.2 lists the attributes of self-Aware, environment-aware, self-monitoring, and self-adjusting, which are necessary in achieving the CHOP objectives. The last branch on this tree lists the approaches that must be undertaken to provide an autonomic solution.

Systems should automatically adapt to dynamic changing environments [5]. This self-configuration objective of self-management enables the introduction of new entities such as new devices, roaming devices, software services, and even personnel, into a system with little or no downtime or human interaction. Configuration will be under the control of well-defined SLA (service level agreements) or high-level directives or goals. When a new entity is introduced, it will integrate itself effortlessly, making its presence and functionality public knowledge while the host system and existing entities reconfigure themselves if necessary.

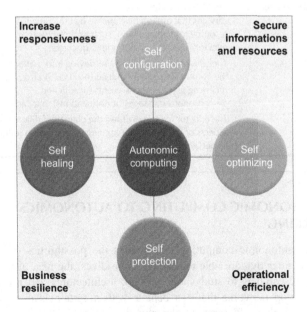

FIGURE 1.2   IBM Autonomic Computing Self-CHOP. By implementing self-CHOP Management, computing systems are themselves able to solve any operational problems related to configuration, healing, optimization, or protection reducing the burden on human administrators.

• Autonomic manager (AM) implements a control loop called MAPE that implements the life cycle from monitoring managed elements and autonomically modifies their state to fulfill the administrator goals (see Table 1.1):

TABLE 1.1

Monitor Function	Allow the AM to collect, aggregate, filter, and report details (e.g., metrics, topologies) from the underlying management element(s) under its responsibility.
Analyze Function	This allows the management element (ME to analyze the collected details to understand the current system state. This analyze function requires the use of complex models of the various situations in which the management elements(s) could evolve.
Plan Function	Once the situation is identified, the ME needs to define the set of actions needed to achieve the high-level goals and objectives.
Execute Function	This function allows the ME to change the behavior of the managed resource using effectors.
Managed Resource	The managed resources are the underlying controlled system components. These can be a server, a router, a cluster or business application, and so on.
Manageability Interface	The manageability interfaces are all the management services that are made available by the managed resource to manage them, such as the sensors and the effectors used by an autonomic manager.
Sensor	The sensor interface allows retrieving information about the current state of a managed resource. It also allows receiving asynchronous events (unsolicited, asynchronous messages or notifications) that can occur.
Effector	The effector interface allows the change of different aspects of the state of the managed resource to influence its behavior.

FROM AUTONOMIC COMPUTING TO AUTONOMICS NETWORKING

Similarly to autonomic computing that studies the possibilities of developing new computer that are able to manage themselves, the idea of autonomic networking research is to study new network architectures and protocols in order to develop networks that can behave with a certain level of freedom and adapt themselves efficiently to new situations (possibly unknown initially) without any direct human intervention [9]. Systems management process can be simplified by automating and distributing the decision-making processes involved in optimizing the system operation. This will save cost, enabling expensive human attention to focus more on business logic and less on low-level device configuration processes.

*"Device, system and network **intelligence** that enable the services and resources offered to be **invisibly adapted** to changing user needs, business goals, and environmental conditions according to **policy** and **context**."*

—John Strassner

It is very important to understand how to build such a dynamic system in which any change in an individual element (networked elements) could affect many others and vice versa.

Therefore the main complexity involved in applying the autonomic concepts to networking lies precisely in understanding how desired behaviors are learned, influenced, or changed, and how, in turn, these affect other elements, groups, and networks. This could concern any aspect of the network or support services such as configuration, performance, security, or routing (see Table 1.1).

An Autonomic Network Elements (A-NE) (Figure 1.3) performs the following tasks:

TABLE 1.2

Sensing its environment	This is similar to the sensors in the autonomic computing architecture. The A-NE should continuously monitor the managed element (s) under its control using different types of sensors that could be software or hardware local or remote. Sensors in this case should be able to intervene at different levels of the communication stack (hardware, protocol, service, application, etc.), which makes it very complex.
Perceiving and analyzing the context	When information is collected from the sensors, the A-NE needs to understand its context. Indeed, the information collected could have different meanings based on the context in which the A-NE is evolving. The network environment is intrinsically very complex. Therefore, it is a very difficult task as the A-NE will need to interpret heterogeneous monitored information using other levels of information (local or global). The information can be related to different aspects of the network or the supported services. Historical information is very important to analysis of the context and understands in which situation the A-NE is now. Based on high-level goals, the A-NE will try to maintain itself in a set of "desired states" according to the context. Otherwise, the A-NE will need to autonomously perform some actions to change its state and try to reach a desired one. The desired state could sometimes be the optimal one, but in some situations, this state could only be

(Continued)

TABLE 1.2 (*Continued*)

	a safe one that would ensure that the system will always deliver the service. This state could then be changed to move toward an "optimal one."
Learning	During its lifetime, the A-NE will face different contexts and situations to which it has reacted, implementing different strategies to always fulfill its assigned goal. During these trials, the A-NE will be able to evaluate the usefulness of the implemented situation and learn from the performed action to adapt itself to future known or unknown situations. Autonomic adaptation is the capability by which the A-NE will improve by learning (increasing its knowledge) the best strategies to plan in order to react to situations. This is what humans do to improve their knowledge and skill but in autonomic networking, this should be an inner capacity of network elements.
Participating in Groups	A-NE cannot improve their knowledge if they do not interact with other A-NE to improve its knowledge and skill. A-NEs need to communicate, collaborate, andexchange information and knowledge to improve their capabilities to solve problem, better their performance, and secure themselves. These group interactions are also important in collectively achieving a global goal that cannot be reached without a certain level of coordination. These communications should be achieved within purposeful (structured and unstructured, ad hoc) groups or clusters. Information should be understandable to the A-NEs, though it is exchanged by autonomic entities.
Planning	Once the context is identified and its situation is evaluated, the A-NE should define the strategy (list of actions) that should be taken to either reach a "desired state" in case it is in an "undesired state" or to reach another "desired state" that is better from a different perspective, that is, performance, security, organization, and the like. Therefore the planning process will encompass a set of strategies that allow the A-NE to continuously fine tune the underlying managed elements and adapt to new contexts while always seeking to be in "desired states." With the distributed nature of the network, the planning can be very difficult as it is not possible to enforce an action instantaneously; when a set of actions are identified, it is not possible to activate them also at the same time. As the A-NEs take their decision in an autonomic way, a consistently among the actions should be ensure in a completely distributed and decentralized way which is a real challenge. Here we seek convergence as actions can be inconsistent, and therefore convergence time becomes an important aspect.
Actuating its state	Finally, the A-Ne should have the full control of itself and the parameters that affect its local behavior. This shall happen through a set of actuators that are linked to the underlying physical and logical resources that are part of the A-NE's boundaries.

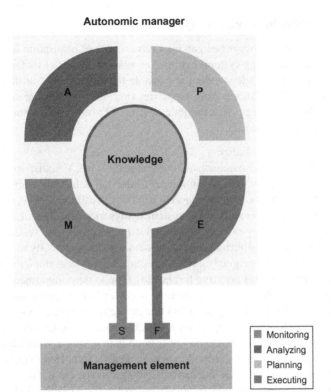

FIGURE 1.3 IBM Autonomic Element Architecture and Inner MAPE Loop. The autonomic element continuously senses the underlying management element state and executes the MAPE loop to identify the appropriate actions to enforce so that the management element is always in the desired state.

AUTONOMIC (NETWORKING) DESIGN PRINCIPLES

Autonomic networking can be built on a set of design principles that have already been proven to provide some level of autonomic behavior in many areas. Many research projects have followed different directions in achieving autonomic systems with different levels of success. Indeed, the principles of autonomic behavior already exist, not only in the natural system (nature, biology), the social environments (society, communities, etc.), but also other areas of IT in fields such as space, vehicular, robotics. These principles can be applied solely or collectively to build the foundation of autonomic networks.

When studying these areas, it is possible to identify some general design principles that can help to build autonomic systems. These principles can also be used in the area of networking, taking into account its particular specificities such as heterogeneity, scalability, and distribution.

Living Systems Inspired Design

Living systems have always been an important source of inspiration for human-designed systems. Living systems exhibit a number of properties that make them autonomic, and their understanding is valuable for the design of artificial autonomic systems [19]. Among these properties two characteristics are especially interesting for the design of autonomic systems (1) bio-inspired survivability and (2) collective behavior.

- Bio-Inspired Survivability
  The body's internal mechanisms continuously work together to maintain essential variables within physiological limits that define the viability zone. In this very complex system, adaptive behavior at different levels is directly linked with the survivability. The system always tries to remain in an equilibrium zone, and if any external or internal stimulus pushes the system outside its physiological equilibrium state the system will autonomically work toward coming back to the original equilibrium state. This system implements what is called positive and negative forces [28]. The system implements several internal mechanisms that continuously work together to maintain essential variables within physiological limits that define the viability zone

  In the biological system, several variables need to be maintained in the viability zone, that is, upper and lower bounds (e.g., sugar level, cholesterol level, body temperature, blood pressure, heart rate). However, environmental change may cause fluctuations (food, efforts, etc.). The autonomic mechanisms in the body continuously control these variables and maintain them in the viability zone (Figure 1.4). This is called homeostatic equilibrium. Three types of adaptation to environmental disturbance are available to higher organisms (see Table 1.3):

TABLE 1.3

Short-term changes	This adaptation allows the biological system to respond to a stimulus immediately: For example, an environmental temperature change moves the body temperature variable to an unacceptable value. This rapidly induces an autonomic response in the (human) organism, that is, either perspiring to dissipate heat or shivering to generate heat. Such adaptation is quickly achieved and reversed.
Somatic changes	Prolonged exposure to environmental temperature change results in the impact of the change being absorbed by the organism, that is, acclimatization. Such change is slower to achieve and is reversible once the individual is no longer in that specific environment.
Genotypic changes	A species adapts to change by shifting the range of some variables, for example, in a cold climate a species may grow thicker fur. Such genotypic change is recorded at a cellular level and becomes hereditary and irreversible in the lifetime of the individual. The adaptation here is through mutation and hence evolution.

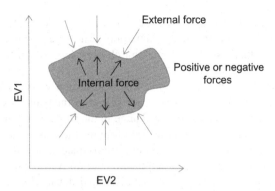

FIGURE 1.4 Biological Physiological Equilibrium. The environment in which the biological system evolves is always pushing it outside its viability zone. The biological systems have inner autonomic physiological mechanisms that create reverse forces to maintain the bio-system in viability. (This equilibrium is called homeostasis equilibrium.)

    These adaptation mechanisms have a primordial role in the biological system's ability to adapt to changes in environment for survivability and evolution toward more sophisticated systems. This can and should inspire the design of future autonomic systems as these intrinsic properties of biological systems could similarly benefit artificial autonomic systems such as autonomic networks [39].

- Collective Behavior

  The term *collective behavior* was coined by Robert E. Park and employed by Herbert Blumer to refer to social processes and events that do not reflect existing social structures (laws, conventions, and institutions), but that emerge in a "spontaneous" way [31]. Among the collective behavior one needs to understand, "social movement" is particularly relevant to understand—notably, an understanding of how from an individual autonomic behavior some general behavior emerges. Social movement is a form of collective behavior that is identified by a type of group action performed by the individuals within the movement. It generally emerges from a large informal grouping of individuals and/or organizations focused on specific political or social issues. The growth of the social movement is sometimes not under any control and emerges because many factors and circumstances take place at the same time. Recently, some modern movements have utilized technology such as cellular networks with the SMS service or the Internet with Twitter to mobilize people globally. From this point of view, it is interesting to notice how local information becomes global and can influence the behavior of all the members of the community. The architecture of autonomic systems can learn a lot from such social collective behavior, which represents a real large-scale prototyping of autonomic entities (humans) interacting to fulfill different and sometimes conflicting goals. In this context, economists and social science researchers have made use of game theory to try to model the behavior of individuals able to take autonomous decisions in strategic situations. The objective is to understand how an individual's success in making choices depends on the choices of others.

The objective is to identify potential equilibrium which is beneficial to all parties. In an equilibrium state, each player of the game has adopted the most appropriate strategy and any change in the strategy will not be beneficial to any of them. Among these equilibriums, the Nash equilibrium is an attempt to capture this idea. These equilibrium concepts are motivated differently depending on the field of application, so do Autonomic Networking. This methodology is not without criticism, and debates continue over the appropriateness of particular equilibrium concepts and the appropriateness of equilibrium altogether in term of fairness. For example, what is the best configuration of a base station to share the radio resources among a number of users attempting to access the network? AN-Es could have to take their own decision in their own situation (local knowledge) and maybe a partial knowledge of what is happening at the global level.

These collective behavior studies can inspire the design and development of novel network architecture that exhibit flexible and dynamic organization. This collective behavior could rise from individual AN-Es interacting and reorganizing among themselves according to some high-level objective such as resource management, topology change, or economic need. Figure 1.5 shows how an individual AN-E seamlessly integrates an already organized network (e.g., ad hoc network, new network equipment in an existing infrastructure network, etc.). The figure also shows how two networks can integrate seamlessly together without any manual intervention.

Policy-Based Design

"Policy is a rule defining a choice in the behavior of a system" [11, 12]. Policy concept has already been widely used in the area of networking to introduce some level of automation in the control and configuration of network equipment behavior based on a set of predefined event-condition-action rules defined by the administrator. A policy is typically described as a deliberate plan of action to guide decisions and achieve rational outcome(s). The human expert defines the right policies, and therefore these policies are enforced in the network [14]. Once enforced, the network could govern itself, freeing the administrators from having to deal with all the known situations and appropriate reactions. Policy-based behavior is very useful but cannot be used by itself because situations can change and it is not always possible to know about all situations and the autonomic network needs to exhibit other properties such as adaptability to varying contexts and situations [15]. Policy-based design permits a level of separation of concerns to facilitate the design of the autonomic system. This will be made feasible by means of behavior definitions and conflict resolution [11–13]. It is thought that a number of *meta-roles* associated with the

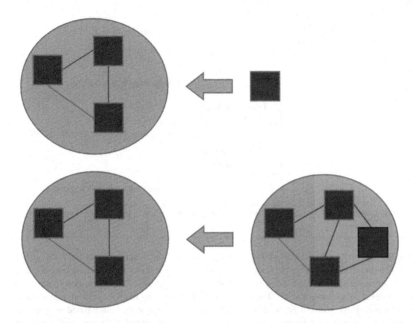

FIGURE 1.5 Autonomic composition et decomposition of Autonomic Networks from individual or group of Autonomic Element(s). Collective behavior in a society helps one to understand how individual interactions lead to more efficient collective behavior. The self-organization and communications mechanisms that are used in this context could inspire the development of autonomic networks exhibiting the same interesting properties.

role's actor, such as role director and role observer, might be needed. Role actors become active when triggered by role-specific headers found in IP datagrams. Role directors are triggered by a need to specify/activate/enforce new behavior at associated role actors [40]; role observers are triggered (as specified in their behaviors) by any of the above events pertaining to the task of monitoring or auditing the behaviors of role actors and directors.

Autonomic networking could be inspired by the policy-based design (Figure 1.6) proposing a proactive approach, in which autonomic networks will self-organize based on these policies [10]. Policies can be used at different levels of the autonomic loop: monitoring, analyzing, planning, and enforcing. Autonomic networking could take advantage of policy monitoring, policy analyzing, policy decision, policy enforcement logical points, and so on, to implement the autonomic loop. Of course, the same conflict problems could rise from using these policies at different decision levels, and this approach will also require defining efficient conflict resolution mechanisms.

Context Awareness Design

Annind Dey defines context as "any information that can be used to characterize the situation of an entity" [17]. Dr Annind K. Dey from Berkey Research Labs defines context "as any information relevant to an interaction that can be used

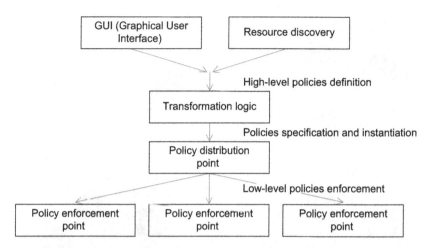

FIGURE 1.6   Policy Based Design. High-level policies are specified by experts and introduced in the system to be enforced in the network elements. To ease the work of the administrator, high-level policy languages are used, and the system introduces several levels of transformation to transform the high-level policies into low-level policies that are distributed by the policy decision point to the different policy enforcement points that control the network elements. Policy-based design gives administrators a way to delegate to the system some level of control of the underlying network elements.

to characterize the situation of an entity: An entity is a person, place or object that is considered relevant to the interaction between a user and an application, including the user and applications themselves" [18]. In general, we can define a context as a set of rules of interrelationship of features in processing any entities (including the entity solely). Computer science has introduced the idea of context awareness in order to develop applications that can sense their environment (localization, time, user identity, etc.) and react accordingly. Chen and Kotz argue that context can be categorized as having two main aspects: active context and passive context [20]. Active context is context information that influences the behaviors of an application, whereas passive context represents information that is relevant but not critical to an application. Gwizdka distinguishes between context that is internal and that which is external to the user. Internal context encompasses work context, while the external context defines the state of the environment. In autonomic networking, J. Strassner presents a more abstract and extensible definition of the context as follows: "The Context of an Entity is the collection of measured and inferred knowledge that describes the state and environment in which an Entity exists or has existed" [19]. Indeed, the idea is that A-NE could take into account the information about the circumstances under which they are able to operate and react accordingly. Context awareness is also sometimes extended to the concept of situation awareness, which aims also to make assumptions about the user's current situation. But this is somehow limiting as even a machine, service, or protocol, and not just a user, can be in a particular situation. From a practical point of view, the main problem with the context awareness design lies in deciding how to model the context in order to capture all the states as well as the semantics of the interactions. For example, Strassner proposes an approach whereby the context of an entity is modeled as a collection of data, information, and knowledge resulting from gathering measurements of and reasoning about that entity. The originality of the approach is the use of modeling patterns and roles to feature an extensible representation of context. The proposed model could then be refined or extended to deal with specific domains.

Self-similarity Design Principle

The term *autonomic computing* also refers to a vast and somewhat tangled hierarchy of natural self-governing systems, many of which consist of myriad interacting self-governing components that in turn comprise large numbers of interacting autonomous, self-governing components at the next level down as introduced by Jeffrey O. Kephart and David M. Chess from IBM [15]. This vision of autonomic computing applies also to networking area however with more complexity than in computing systems due to the distributed, large scale and heterogeneous nature of networks. Interacting autonomic element need to organize among themselves so that some system level properties autonomically emerge from that organization and make the whole system autonomic (Figure 1.7). Such a scalable organization can also be found in what is called self-similarity. This design principle suggests that the same organization will be found at different scales of the system. This allows using the same template to create higher level templates having the same properties such as ice snowflakes or glass cracks. The same principle can be used to design the autonomic system at a small scale, which is then able to grow to large scale while maintaining small-scale properties. This will simplify the design of the system and solve the

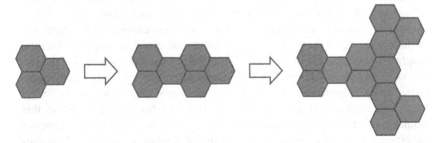

FIGURE 1.7 Self-Similarity as a Design Principle for Autonomic Networks. The natural and biological systems have shown very interesting self-similarity properties allowing them to grow in a structured way and exhibiting scalability properties. Scalability is an important problem in autonomic systems that is difficult to reach and sometimes to prove. Inspiring from self-similarity properties of exiting natural systems could help to design an autonomic system with the required properties.

problem of scalability, which is one of the main issues involving the Internet today.

Self-similarity will also permit achieving self-organization in the autonomic system. S. Camazine [23] defines self-organization as "a process in which pattern at the global level of a system emerges solely from numerous interactions among the lower-level components of the system." In this self-organization schema, autonomic entities have access only to local information without reference to the global pattern. Still, the system as a whole is able to achieve a global pattern. In the autonomic network context, this aims at designing autonomic network elements that are able to organize among themselves using their own internal processes and not relying on any global entities to control them. This self-organization is the only guarantee of the same properties exhibited at the large scale as those exhibited at a smaller scale. These could be any aspect of interest in networking such as performance, security, and reliability.

Adaptive Design

By definition an autonomic network should be adaptive—that is, be able to change its operations, state, and function to cope with situations that could affect the general goals for which it was built. A-NE should be designed so that their inner behavior can adapt to various situations.

According to Laddaga, self-adaptive software may be defined as "software that evaluates and changes its own behavior when the evaluation indicates that it has not [been] accomplishing what it is intended to do, or when better functionality or performance is possible" [26]. The self-adaptive approach presents an attractive concept for developing self-governing systems that partially or fully accommodate their own management and adaptation activities. Self-adaptation in the design of A-NE can be fulfilled in a local or coordinate way but should always seek to put the global network in a desired state. The adaptation capabilities of each A-NE should be able to deal with temporal and spatial changes (e.g., topology change, structural change), operational changes (e.g., malicious attack, faults), and strategic changes (e.g., objectives of the operators, SLAs). The temporal adaptation could be inspired from the adaptation mechanisms in biology as described in the bio-inspired design. The Autonomic Computing approach suggests an adaptation inspired from the automatic theory to implement a first and fast adaptation loop called the automatic loop. This approach is also applied to the area of networking, however this requires another level of adaptation called cognitive loop that aims to improve the efficiency of the first loop in fully distributed manner over time. This suggests that the A-NE will improve its inner capacity to react to situational changes by learning in the long term from the experiences while with the automatic loop; it is capable of responding to immediate situations enforcing the best actions with the available knowledge. As its knowledge increases in time, A-NE's skill to respond efficiently to various known and unknown situations will also increase.

Knowledge-Based Design

The knowledge plane for the entire chapter is a high-level model of what the network is supposed to do. Some researchers argue that there can be no autonomic network without building an efficient and complete knowledge plane able to capture all the properties of underlying network, protocols, and supported services. However, the heterogeneity, complexity of interacting, and manipulation underlying technology needs to disappear from the user's perspective. The so-called Future Internet is based on the knowledge plane concept, which is a kind of meta-control plane for future intelligent management of the Internet [16]. The knowledge plane was originally proposed as a research objective designed to build "a fundamentally different sort of network that can assemble itself given high level instructions, organize itself to fulfill new requirements change, automatically discover new autonomic elements and integrate them seamlessly when necessary and automatically tries to fix detected problems" [33]. The used term plane comes from the fact that network functionalities are organized into layers (Figure 1.8), each layer is responsible for the management of its own data or information. The knowledge plan come above these layers to aggregate and give semantic meaning to all underlying information as well to any new knowledge that helps the network to fulfill its objectives. Hence, the knowledge plane was envisioned as "a new construct that builds and maintains high-level models of what the network is supposed to do, in order to provide services and advice to other elements of the network" [37]. This approach advocated the use of cognitive and artificial intelligence (AI) techniques to achieve the above goals.

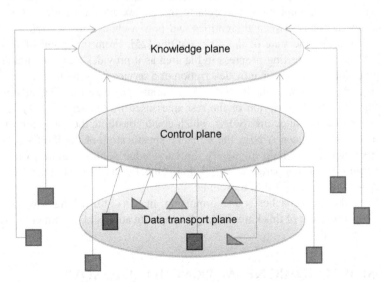

FIGURE 1.8 Knowledge Plane: The knowledge plane is a high-level model of what the network is supposed to do. It constitutes a lingua franca build on information models and ontologies, which can serve as a lexicon to translate between different heterogeneous control interfaces. It constitutes a meta-control plane for future autonomic networks.

The knowledge plane is based on three principal mechanisms: a new architectural construct that is separate from the existing data and control planes; the ability to harmonize different needs; and a cognitive framework (giving it the ability to make decisions in the presence of partial or conflicting information).

Although the knowledge plane approach appears to be very promising, many issues need to be addressed to build autonomic networks. Indeed, existing contributions on knowledge plane do not define how to express business goals and how to link these goals to underlying services and resources. While existing data models are excellent at representing facts, they do not have any inherent mechanisms for representing semantics, which are required to reason about those facts [29]. To build such as knowledge plane, it is necessary to specify how to work with heterogeneous technologies and devices. Several ontology-based approaches have been proposed to solve this problem of integrating heterogeneous domains and to dynamically discover the services and capabilities of autonomic entities. Ontology provides a vocabulary of classes and relations to describe a domain aiming at representing knowledge and sharing it. First, the Web Service Definition Language (WSDL), a standardized service description language of the World Wide Web Consortium, has been used to describe the functional aspects of services by defining the semantics of their input and output parameters, but it remains limited as the internal semantic of the service is not described. For that, OWL-based Web Service Ontology (OWL-S) is being positioned as a good candidate to enhance the descriptions of these services using ontology-based semantics. In general, the Web Ontology Language (OWL) is considered as a candidate for knowledge representation in autonomic ystems and more particularly the contextual information as well as the interactions between the entities and how a change in the state of an entity will affect the state of the other entities [25]. From this point of view, OWL-S marks interesting progress in the area as it provides an unambiguous, computer-interpretable semantic description of a service by providing rich definitions of the semantic of the IOPE (Inputs, outputs, preconditions, and effects), in addition to the description of the resources used by the service [24]. In a service-oriented architecture (SOA), which could constitute the foundation for the interaction between A-NEs, the semantic specification of the IOPEs of the A-NE services could help to specify the knowledge A-NEs require to interact, compose, and cooperate to fulfill their global goal. This should also be associated with some level of reasoning.

While these approaches sound promising, there is still a long way to go before the objective of this knowledge plane specification and instrumentations can be achieved.

FROM AUTONOMIC NETWORKING TO AUTONOMIC NETWORK MANAGEMENT

Autonomic network management is one facet of autonomic networking that focuses on developing new solutions to allow the networks to self-manage.

Traditional network management solutions called Simple Network Management Protocol and Common Management Information Service/Protocol have shown limitations with regard to the increased scale and complexity of existing networks as the intelligence of solving the problems was always outside the network and usually humancentric. Policy-based management solutions have certainly provided a certain level of simplification by automating some aspects of management, but not enough to cope with the ever increasing complexity. The objective of autonomic network management is to investigate how to design new management solutions that will be able to cope with the increasingly complex, heterogeneous scalability of today's and future networks. The solution should benefit from the autonomic concepts to reach the required flexibility and adaptability to deal with any unforeseen situation. The idea behind the autonomic network management solutions is to develop management systems that are capable of self-governing and reducing the duties of the human operators who are not able to deal with increasingly complex situations. The systems should exhibit some level of intelligence so that their capability can improve over time, assuming more and more tasks that are initially allocated to skilled administrators. Humans will only need to interact with the system using some high-level goal-oriented language and not any low-level commands as is true today. This autonomic management of the networks and services will not only improve the end users' quality of service, as problems and quality degradation will be solved much quickly, but it will also reduce operational expenditure for network operators.

As presented earlier, the autonomic network as well as autonomic network management could make use of different techniques to exhibit the required properties. In autonomic network management, human management goals should be dynamically mapped to enforceable policies across the A-NE across the network. A-NEs should exhibit autonomic behavior in term of adaptation to changing context, improving at each stage their capacity to find better solutions. Some research entities think that these adaptations should be constrained by some human-specified goals and constraints, while others think that the emerging behaviors and collective behaviors will freely reach optimum equilibrium without any human intervention.

From an operator's point of view, the full freedom of the network is difficult to accept today, so "self-management" would be appropriate only if it were to be overseen or governed in a manner understandable to a human controller. Experience with the Internet in the past has shown that several visions could coexist.

Autonomic network management presents many challenges that need to be addressed. Among these challenges, the smooth migration from existing management to fully autonomic network management will require an accurate mapping between underlying data models and high-level semantic models in order to efficiently control the underlying heterogeneous network equipments and communication protocols (Figure 1.9). On the reverse side, a high-level governance directive should also be correctly mapped down to low-level adaptation and

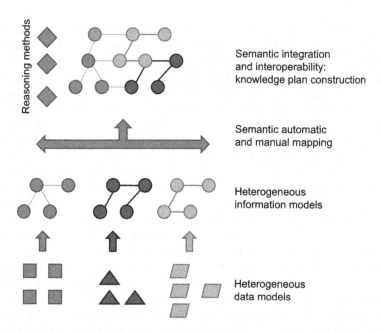

Reasoning methods

Semantic integration
and interoperability:
knowledge plan construction

Semantic automatic
and manual mapping

Heterogeneous
information models

Heterogeneous
data models

FIGURE 1.9   Integration of Heterogeneous Underlying Management Information Sources. Autonomic network management cannot be achieved without a unified knowledge plan gathering all
the knowledge about the underlying possibility of the complex and heterogeneous network elements to control. A high-level semantic model (sometime called knowledge plan) can be built based
on a transformation process of models starting from the underlying data models up to the knowledge plan. Autonomic Network Management can then be achieved by defining appropriate Self-*
mechanisms that implement the MAPE loop using this knowledge plan.

control policies to be enforced in the individual heterogeneous elements. This
mapping can be even more complicated when one takes into account the context
of a specific service chain or flow within a more richly connected network of
managed components [30].

At the higher level of its architecture, the A-NE should maintain a knowledge base that should help to describe its situation and to reason about it to
determine the right action to perform. Therefore, the A-NE should maintain
different types of knowledge. Which knowledge to maintain, how to represent it, and how to reason on it are among the challenges existing and future
research initiatives will address. The knowledge can be structured in different
levels of abstraction [34]. The following table presents a decomposition of the
knowledge in three layers: domain knowledge, control knowledge, and problem
determination knowledge (Table 1.4). Domain knowledge provides a view or
conceptualization of the managed objects, their properties, the relations among
them, and the like; control knowledge represents the ways to manage and control
the autonomic elements of the domain; and problem determination knowledge
contains all the knowledge necessary to analyze and infer about situations to

find appropriate solutions and describes the problems related to the domain and their corresponding applied solutions.

TABLE 1.4

Domain Knowledge	• Captures knowledge about all the elements and concepts of the target domain. • Contains the knowledge about all the existing managed elements, how to install, configure, and control them • Relations between the different concepts.
Control Knowledge	• Aims to determine whether or not changes need to be made in the managed elements by means of policies for example. • Provides a uniform and neutral way to define the control policies to govern the decision-making process of the autonomic network.
Problem Determination Knowledge	• Represents the information captured or inferred. • Allows defining context and reason on it. • Specifies how to infer new knowledge from existing knowledge to detect symptoms and take decision. • Allows the system to learn about situations and improve its capabilities. • Defines a uniform approach to represent problems and associated potential solutions.

The specification of a knowledge plane for autonomic network management as well as a general architecture is not an easy task. Many initiatives have been launched to address this issue during the last decade, but there has been no agreement so far either on a standard specification of this knowledge plan or on a common architecture. It is not sufficient to find the right mapping techniques, but it is also very important to agree on the structure of the common representation of knowledge. As stated in [38], current approaches for building an Network-Knowledge-Base System (NKBS) are mainly limited to representing the network structure knowledge with some efforts to build simple models for control knowledge.

CONCLUSION

This chapter has presented a general overview of the autonomic concept as applied to networking and to network management. It highlights the different challenges faced by academia and industry. Despite the numerous efforts made in this area, as yet there is no agreement on either the architecture of autonomic networks or autonomic network management, nor is there any consensus on the knowledge plane. Standardization on this critical issue has not really started despite several initiatives with organizations such as IEEE and ETSI. This can be explained primarily by the fact that the autonomic concept has led to the introduction of fundamental changes in the way networks will be built and controlled; therefore the community has not yet reached agreement on the level of autonomic and determinism in the behavior of autonomic element. It is envisioned that some solutions will come from the industry such as those initiated by IBM in autonomic computing and if the experience is positive perhaps it will be deployed on a small scale before being deployed globally. In any case, standardization does not affect all aspects of autonomic networking or network management but rather probably only the architecture and structure of the knowledge plan. The internal mechanisms of autonomic network elements will not have to be standardized and should be part of the competition process between the concerned actors. Nevertheless, the standardization of the communication protocols to enable autonomic networks element to exchange knowledge is mandatory, allowing the autonomic network element to provide autonomic open interfaces to facilitate the interaction between autonomic network elements from different sources.

REFERENCES

[1] Steve R. White, James E. Hanson, Ian Whalley, David M. Chess, and Jeffrey O. Kephart, "An Architectural Approach to Autonomic Computing," Proceedings of the International Conference on Autonomic Computing (ICAC'04), November 2004.

[2] J. Simpson and E. Weiner (eds.), Oxford English Dictionary. Oxford, Oxford University Press, 1989.

[3] Paul Horn; "Autonomic Computing: IBM's Perspective on the State of Information Technology," Technical Report, IBM Corporation, October 15, 2001 http://www.research.ibm.com/autonomic/manifesto/autonomic_computing.pdf

[4] A.G. Ganek, "The dawning of the computing era," IBM Systems Journal, Vol. 42, 2003.

[5] IBM White Paper, An architectural blueprint for autonomic computing. IBM, 2003.

[6] R. Sterritt, "Autonomic computing—A means of achieving dependability," presented at 10th IEEE International Conference and Workshop on the Engineering of Computer Based Systems (ECBS), 2003.

[7] R. Sterritt, "A concise introduction to autonomic computing," Journal of Advanced Engineering Informatics, Engineering Applications of Artificial Intelligence, Vol. 19, pp. 181–187, 2005.

[8] IBM developer works, Can you CHOP up autonomic computing? June 2005. http://www.128.ibm.com/developperworks/autonomic/library/ac-edge4

[9] B. Melcher, "Towards an autonomic framework: Self-configuring network services and developing autonomic applications," Inter® Technology Journal, Vol. 8, pp. 279–290, 2004.

[10] "Autonomic communication," Research Agenda for a New Communication Paradigm, Fraunhofer Fokus, Autonomic Communication, White Paper, November 2004.

[11] E. Lupu and M. Sloman, "Conflicts in policy-based distributed systems management," IEEE Transaction on Software Engineering—Special Issue on Inconsistency Management, Vol. 25, No. 6, November 1999.

[12] J. Moffeett and M. Sloman, "Policy hierarchies for distributed systems management," *IEEE Journal on Selected Areas in Communications,* Vol. 11, No. 9, December 1993.

[13] I. Aib, N. Agoulmine, M. Sergio Fonseca, and G. Pujolle, "Analysis of policy management models and specification languages," IFIP International Conference on Network Control and Engineering for QoS, Security and Mobility, Kluwer, pp. 26–50, October 2003, Muscat, Oman.

[14] Dinesh C. Verma and D.C. Verma, *Policy-Based Networking: Architecture and Algorithms,* New Riders Publishing, Thousand Oaks, CA, 2000.

[15] J.O. Kephart and D.M. Chess, "The vision of autonomic computing," *IEEE Computer Magazine,* Vol. 36, 2003, pp. 41–45. http://www.research.ibm.com/autonomic/research/papers/AC_Vision_Computer_Jan_2003.pdf

[16] D. Clark, C. Partridge, Chs. Ramming, and J. Wroclawski, "A knowledge plane for the Internet," ACM Sigcomm 2003, Karlsruhe, Germany, August 2003.

[17] A. Dey, "Providing architectural support for building context aware applications," Ph.D. Thesis, 2000.

[18] A.K. Dey, G.D. Abowd, and A. Wood, "CyberDesk: A framework for providing self–integrating context–aware services. Knowledge based systems," Vol. 11, No. 1, pp. 3–13, September 1998.

[19] J. Strassner et al., "Modelling context for autonomic networking," IEEE 2008.

[20] G. Chen and D. Kotz, "Solar: A pervasive-computing infrastructure for context aware mobile applications," Dartmouth Computer Science Technical Report TR2002-421, 2002.

[21] J. Gwizdka, "What's in the context?" Position paper for workshop on The What, Who, Where, When, Why, and How of Context-Awareness. CHI'2000, 2000.

[22] Martin Feeney, "Autonomic management of ubiquitous computing environments," Master's Thesis, Waterford Institute of Technology, 2006.

[23] S. Camazine, J.-L. Deneubourg, Nigel R.F., J. Sneyd, G. Téraulaz, and E. Bonabeau, Self-Organisation in biological systems. Princeton Studies in Complexity. Princeton University Press, 2001.

[24] J. Keeney, K. Carey, D. Lewis, D. O'Sullivan, and V. Wade, Ontology-based Semantics for Composable of Autonomic Elements. Workshop on AI in Autonomic Communications at 19th International Joint Conference on Artificial Intelligence, JCAI'05, Edinburgh, Scotland, July 30–August 5, 2005.

[25] Yechiam Yemini, Member, IEEE, Alexander V. Konstantinou, and Danilo Florissi, "NESTOR: An architecture for network self-management and organization," *IEEE Journal on Selected Areas in Communications,* Vol. 18, No. 5, May 2000.

[26] R. Laddaga, Active Software, in 1st International Workshop on Self Adaptive Software (IWSAS2000). Oxford, Springer-Verlag 2000.

[27] G. Di Marzo Serugendo, "Autonomous systems with emergent behaviour," chapter in *Handbook of Research on Nature Inspired Computing for Economy and Management,* Jean-Philippe Rennard (ed.), Idea Group, Hershey, PA, pp. 429–443, September 2006.

[28] G. Di Marzo Serugendo, M.-P. Gleizes, and A. Karageorgos, "Self-organisation and emer-gence in MAS: an overview," *Informatica* 30(1): 45–54, Slovene Society Informatika, Ljubljana, Slovenia, 2006.

[29] L. Stojanovic et al., "The role of ontologies in autonomic computing systems," *IBM Systems Journal,* Vol. 43, No. 3, 2004.

[30] Towards autonomic management of communications networks Jennings, Van Der Meer, Balasubramaniam, Botvich, Foghlu, Donnelly, and Strassner, *Communication Magazine,* 2007.

[31] H. Blumer, Collective behavior chapter in *"Review of Sociology—Analysis of a Decade,"* J.B. Gittler (ed.), John Wiley & Son, 1954.

[32] J. Strassner, N. Agoulmine, and E. Lehtihet, "FOCALE: A novel autonomic computing architecture," LAACS, 2006.

[33] D. Clark, C. Partridge, J. Ramming, and J. Wroclawski, "A knowledge plane for the Internet," Proceedings ACM SIGCOMM, 2003.

[34] D. Clark, "The design philosophy of the DARPA Internet protocols," Proceedings ACM SIGCOMM, 1988.

[35] J. Saltzer, D. Reed, and D. Clark, "End-to-end arguments in systems design," Second International Conference on Distributed Systems, 1981.

[36] D. Clark et al., "NewArch: Future generation Internet architecture," Final technical report.

[37] J. Strassner, M.Ó. Foghlú, W. Donnelly, and N. Agoulmine, "Beyond the knowledge plane: An inference plane to support the next generation internet," 2007 1st International Global Information Infrastructure Symposium, GIIS 2007—"Closing the Digital Divide," pp. 112–119, Marrakech, Morroco, 2007.

[38] N. Samaan and A. Karmouch, "Towards autonomic network management: An analysis of current and future research directions." This paper appears in: *Communications Surveys & Tutorials,* IEEE, Vol. 11, Issue 3, pp. 2–35, ISSN: 1553-877X, 2009.

[39] S. Balasubramaniam, S. Botvich, N. Agoulmine, and W. Donnelly, "A multi-layered approach towards achieving survivability in autonomic network," Proceeding—2007 IEEE International Conference on Telecommunications and Malaysia International Conference on Communications, ICT-MICC 2007, pp. 360–365, Penang, Malaysia, 2007.

[40] C. Stergiou and G. Arys, "A policy based framework for software agents," Proceedings of the 16th international conference on Developments in applied artificial intelligence table of contents, Laughborough, UK, pp. 426–436, 2003.

Chapter 2

Autonomic Overlay Network Architecture

Ibrahim Aloqily and Ahmed Karmouch

INTRODUCTION

The growth of the Internet in terms of size and speed, as well as the flood of network applications and services that have been deployed in the last few years, is indicative of a shift from the traditional communication systems designed for simple data transfer applications to highly distributed and dynamic systems. Naturally, the spread of such systems has led to an increase in multimedia development, which in itself is a feature that has become indispensable in networking environments. Audio and video content on the Internet is more popular than ever, and many systems are designed with the purpose of carrying this media; video conferencing, video on demand, IP Telephony, and Internet TV are but a few. In addition to being of large scale, these distributed networks and applications are unpredictable and complex; they are highly dynamic in changing environments. As a result, their management (networks and applications) is continuously faced with new complexities, putting the burden on the shoulders of network managers and service providers to design and implement mechanisms that recognize the nature of different applications demands, and that can conform to various users' requirements. This has left management system paradigms in a continuous struggle to keep up with the ever increasing demands and with advancing technologies.

Another factor that has contributed to increased management complexity is the rapid growth of overlay networks and their users. Overlay networks consist of a set of nodes that are connected via virtual links, and are built on top of other computer networks with the purpose of implementing new applications that are not readily available in the underlying network. They can be used to increase routing robustness and security, reduce duplicate messages, and provide new services for mobile users. They can also be incrementally deployed on end hosts without the involvement of Internet Service Providers (ISPs), and they do not incur new equipments or modifications to existing software or protocols. Overlay networks are becoming more popular because of their flexibility

Autonomic Network Management Principles. DOI: 10.1016/B978-0-12-382190-4-00002-4

and their ability to offer new services; extensive research that has been recently exerted in the realms of overlay networks has focused on the design of specific networks to deliver media in a heterogeneous environment. In that course, a specific overlay network for each multimedia delivery service is created, leading to hundreds of overlays coexisting, and as a result, increasing management complexity and posing additional challenges to ISPs. This—in addition to rapid growth of systems such as P2P networks, pervasive computing networks, wireless sensor networks, ad hoc networks, and wireless communication technology—renders traditional network management operations insufficient and incurs new requirements on the networks: to become autonomous, scalable, interoperable, and adaptable to the increasingly dynamic and the widely distributed network demands.

Management involves the tasks of planning, allocating, configuring, deploying, administering, and maximizing the utilization of the underlying network resources. Functionalities of a management system also include aspects such as authorization, security management, reliability assurance, and performance guarantees. Little progress has been made in addressing the problem of designing an overall autonomous management framework for service-specific overlay networks that can be self-configurable and adaptable by automating their management tasks.

Smart Media Routing and Transport (SMART) architecture [1] (as explained later in this chapter) has been proposed to enable the seamless integration of next-generation multimedia services into intelligent networks. It is a multimedia delivery method that enables the flexible configuration of virtual networks on top of the underlying physical network infrastructure. Virtual networks in SMART are deployed for every media delivery service (or group of services), which allows for the configuration of appropriate, high-level routing paths that meet the exact requirements (e.g., QoS, media formats, responsiveness, cost, resilience, or security) of a media service. This has been realized using the concept of Service-Specific Overlay Networks (SSONs). As we will explain, SSONs have the ability to customize the virtual network topology, and the addressing as well as the routing at the overlay level according to the specific requirements of a media delivery service. In addition to that, SSONs transparently include network-side functions into the end-to-end communication path from the MediaServer (MS) to the MediaClient (MC), thus making it possible to support media routing, distribution, adaptation, and caching over complex communication mechanisms such as peer-to-peer communication, multicasting, and broadcasting.

However, SMART does not specify the means by which SSONs are constructed and managed. Creating an SSON for each media delivery session implies that a numerous number of SSONs will coexist and thus, if left unmanaged, they will not only degrade the performance of each other, but also that of the underlying network. In addition, it is essential to have suitable mechanisms to discover the required media processing functions and to seamlessly integrate them in the multimedia delivery session. Moreover, once SSONs are created, there should be a mechanism to adapt them dynamically to the ever-changing conditions of the network, users, and service providers.

RELATED WORK

Requirements posed by autonomic overlay management cause certain problems to emerge; our architecture proposes to resolve these problems, but we still need to characterize them, and that is what we do before presenting our solution. To achieve that objective, in this section we present a survey of current research efforts related to overlay management.

Automated Management for Overlay Networks

Current approaches in the literature present simple adaptation algorithms that offer suboptimal solutions to the management problem. Dynamic self-adaptation in response to changing QoS needs; resources availability; service cost; and perceived performance of the network components, or even neighboring networks, will become an essential operation in future networks. In the following, we investigate some of the few trials for automating one or more of the overlay network management functionalities.

The CADENUS (Creation and Deployment of End-User Services in Premium IP Networks) project [2] attempts to automate network service delivery. The focus was on how QoS technologies can be controlled and managed via standard interfaces in order to create, customize, and support communication services for demanding applications. Mediation components are used to represent the main actors involved, namely users, service providers, and network providers, and define their automated interactions. By defining roles, responsibilities, and interfaces, the service deployment process is decomposed into a set of subprocesses whose mutual interactions are standardized. The model brings novel contributions to automated management. Nevertheless, it lacks scalability and does not discuss the impacts of network heterogeneity on system performance.

The DHARMA (Dynamic Hierarchical Addressing, Routing, and Naming Architecture) [3] proposes a middleware that puts no constraint on the topologies of the overlays and defines a distributed addressing mechanism to properly route data packets inside the overlay. It separates the naming and addressing of overlay nodes, and so it can be used to enable network applications to work over the Internet in an end-to-end mode while seamlessly exhibiting mobility, multicasting, and security. The routing is greedy and follows the closest hierarchy to the destination node. The middleware achieves reasonable results for network dynamics <= 10% and restricts overlays to end-to-end communications.

The Autonomous Decentralized Community Communication System (ADCCS) [4], [5] provides a framework for large-scale information systems, such as content delivery systems. It forms a community of individual members having the same interests and demands at specified times. It allows the members to mutually cooperate and share information without loading up any single node excessively, and it organizes the community network into multilevels of subcommunities. ADCCS is concerned with reducing both the communication delay of a message that is broadcast to all community nodes (while considering latency among them), and the required time for membership management.

In [6], a distributed binning scheme is proposed to improve routing performance by ensuring that the application-level connectivity is harmonious with the underlying IP-level network topology. In the binning scheme, nodes partition themselves into bins such that those nodes that fall within a given bin are relatively close to one another in terms of network latency. To achieve this, a set of well-known landmark machines are used and spread across the Internet. An overlay node measures its distance, that is, round-trip time, to this set of well-known landmarks and independently selects a particular bin based on these measurements. The scheme is targeted at applications where exact topological information is not needed, such as overlay construction and server selection; however, it provides no support for the application-specific demands.

In [7], [8], and [9], a social-based overlay for peer-to-peer networks is proposed. The social-based overlay clusters peers who have similar preferences for multimedia content. A similarity between two peers exists if both share common interests in specific types of multimedia content; hence peers sharing similar interests can be connected by shorter paths so that they can exchange multimedia content efficiently. Specifically, whenever a peer requests an object of interest, it can locate the object among its neighboring peers—that is, the peers that have high similarity and that are more likely to hold the requested object. Some of these approaches [7] model a distance measure that quantifies the similarity between peers, and use the random walk technique to sample the population and discover similar peers from the randomly selected samples. In [8], the similarity of peers is measured by comparing their preference lists, which record the number of the most recently downloaded objects. However, a new user who has only made a few downloads cannot get an accurate similarity measure. In [9], a central server collects the description vectors of all users and establishes overlay links based on the distance between each pair of users. The central server does not explicitly define the description vector, however, which has a significant effect on the accuracy of the similarity measure.

Autonomic Management

Autonomic computing (AC), launched by IBM in 2001 [10], is an emerging technology designed to allow users to traverse transparently and dynamically between different providers and service domains. IBM identified the complexity of current computing systems as a major barrier to its growth [10], and as a result, automated selection of service configuration, relocation, and monitoring must be carried out with minor intervention of users and system administrators. Although, in theory, AC seems to provide the ultimate solution for the complex management problem, in general, research efforts toward autonomic management are still in their infancy and still face many challenges.

In autonomic overlays we focus on automating the management of SSONs; thus, the interaction between the network and computing entities is based on a service request/offer concept in which each entity is responsible for its internal state and resources. An entity may offer a service to other entities. The offering entity responds to a request based on its willingness to provide a service in its current state. Our work is concerned with all possible phases of the service delivery in SSONs—from the instance of requesting a service to terminating it. As a result, we present an integral approach to service providers (SPs) that wish to deliver services over their infrastructure.

SMART MEDIA ROUTING AND TRANSPORT (SMART)

Our lab (IMAGINE) was involved in the European project, the Ambient Networks [11], in which a working group has developed a subproject called Smart Media Routing and Transport (SMART) [1]. The work presented in this chapter was developed using SMART as the starting point. The overall goal of the Ambient Networks Integrated Project [13] is to develop a vision for future wireless and mobile networks. The aim of this project is to create an innovative, industrially exploitable new internetworking framework that is based on the dynamic composition of networks. A key aspect of the project is to establish a common control layer for various network types, which provides end users with seamless multi-access connectivity to enable selection of the best available network. For an operator, the ambient network concept allows flexible and dynamic network configuration and management.

In the environment targeted by Ambient Networks, there will be a broad heterogeneity of access networks, terminals, network interfaces, users, signaling, and transport protocols, applications, and services. As a consequence, certain independent streams of multimedia data may be required to be proactively cached, transcoded, split, synchronized, translated, filtered, legally tapped, or transformed in some way or another before they can be delivered according to a variety of constraints, or properly displayed to the user. To this end, Smart Media Routing and Transport (SMART) architecture [1] has been proposed to enable the seamless integration of next-generation multimedia services into Ambient Networks.

Media Processing Functions

Services, as defined in the SMART context, can be simple requests of information (Web browsing), multimedia streaming (audio and video), and/or conferencing, or they can be more complex service scenarios including mobility features, media adaptation features, and caching features. Media that are delivered as part of SMART-like services may need to be processed along the media path and thus inside the network (e.g., dynamic transcoding of video and audio streams to adapt to changing link properties, or proactive smart caching following user movement). Since services such as media adaptation and transcoding can only be located at the end systems today, they are often of very limited value. In the case of server-side adaptation, the media have to be transmitted several times (once for each type of encoding). Client side adaptation, on the other hand, has the drawbacks of wasting network resources (as the downscaling of the media format is done only at the client end) and increasing the complexity (and hence the cost) of user terminals.

Other services, such as caching or optimal routing of media traffic in order to optimize the possible achievable QoS, can only be achieved using network-side intelligence. Similar reasoning can be used for broadcasting and multiparty communication. Only with the help of network-side components is it possible to optimize the bandwidth usage. Therefore with SMART, additional intelligence can be located at the provider and inside the network. Examples of such intelligence include the following features: media routing and media adaptation to deal with terminal and user mobility; media splitting to enable session/flow mobility; synchronization for recombining split flows; smart caching for accommodating low-bandwidth access networks.

In SMART, multimedia transformation is carried out by network-side media processing capabilities and transformation services [14], termed MediaPorts (or MPs), which are located somewhere on the media path, between the sink, called MediaClient (or MC) and the source, called MediaServer (or MS). MPs must be able to transform the multimedia data originating from the MS into a form that is acceptable for the MC.

Overlay Routing

SMART promotes the concept of overlay networks in order to enable inclusion of the above-mentioned media processing functions in the end-to-end media delivery path in a way that is transparent to the underlying network (i.e., without the need to replace the existing infrastructure) as well as to the end-user applications. Consequently, the migration path from legacy networks toward SMART-enabled Ambient Networks is expected to be inexpensive and straightforward. One important advantage of the overlay concept is that it enables the establishment of different types of overlay networks as needed. This allows, for example, for tailoring the virtual addressing scheme and the overlay routing to best suit the requirements of a particular service. Another example of the tremendous capabilities of overlay routing includes more advanced multimedia transport techniques that enable transparent integration of value-added media processing capabilities into the end-to-end media delivery path. Because of such advantages, the overlay concept has been selected as the basic building block for the SMART framework.

Service-Specific Overlay Networks

A Service-Specific Overlay Network (SSON) is defined through the set of Overlay Nodes (ONodes) that are part of a particular service (or collection of services that are combined into one composed service) and the virtual links that connect the individual ONodes to each other. In SMART, a different virtual network is deployed for every media delivery service (or group of services), which allows for the configuration of appropriate, high-level routing paths that meet the exact requirements (for example, QoS, media formats, responsiveness, cost, resilience, or security) of a media service. Moreover, the exploitation of overlay network techniques also facilitates the transparent inclusion of network-side media processing functionalities (such as caching, adaptation, and synchronization) into the end-to-end data paths. Moreover, the overlay network is able to react dynamically to a changing environment; that is, modifications in the overlay might be triggered due to changes in user preferences, mobility, QoS, or the underlying network. Finally, to provide maximum flexibility, SMART supports all these actions separately for each flow of the media service within a SSON.

Figure 2.1 (redrawn from [1]) illustrates how the SMART architecture relates to the overall Ambient Network architecture. The figure also shows the Ambient Control Space (ACS) as well as its control interfaces, namely, the Ambient Service Interface (ASI) and the Ambient Resource Interface (ARI).

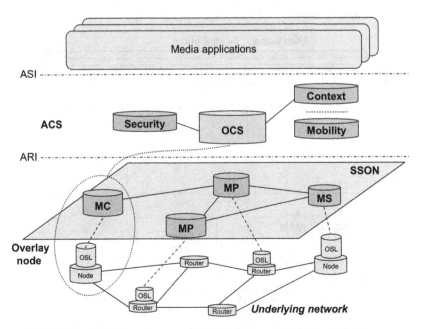

FIGURE 2.1 SMART Architecture within the Overall Ambient Networks Architecture.

Roughly, the ASI provides the service and user profile to the Overlay Control Space (OCS) in case of a request for a media delivery service. The ARI is the interface to the connectivity layer and manages the underlying connectivity resources.

Overlay Node (ONode) Architecture

An ONode is a specialized Ambient Network node that implements the functionality required to join the SSONs by, for example, provisioning network-side media processing functionalities, such as caching, media adaptation, synchronization, and Media aware inside the network. ONodes (see Figure 2.2, redrawn from [1]) can be described from the user perspective and the control perspective. For each SSON of which the ONode is part of, MediaPorts (MPs) are instantiated. MPs are responsible for Media Routing in the control plane and, in the user plane, host the so-called application modules, each of which is responsible for a particular network-side media processing functionality. Furthermore, and depending on the required media processing functionality, overlay nodes can take one or more of the roles of MC, MS, and MP. Note that a physical ONode can be part of many SSONs at the same time.

The *control plane* of the ONode includes the ONode Control entity, which is responsible for the general management of the ONode and the signaling

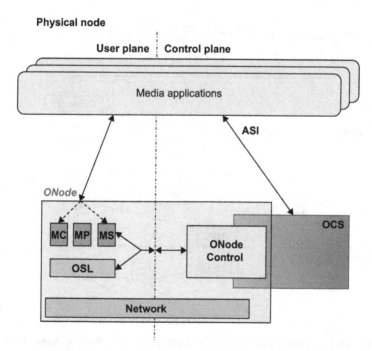

FIGURE 2.2 Implementation of an ONode on a Physical Node.

exchange. The ONode Control consists of several components, which can be classified into those that deal only with the local control and management of the ONode and those that logically belong to the OCS, which is the functional entity residing in the ACS that controls the SSONs on an Ambient Network-wide basis.

The *user plane* of the ONode encompasses the Overlay Support Layer (OSL) and the application modules that take part in media processing actions. The OSL sits on top of the underlying network; it embodies the basic overlay network functionality required in every ONode for handling packets at the overlay level. As such, the OSL is responsible for the sending, receiving, and forwarding of SSON-level packets. The OSL provides a common communication abstraction (overlay level network protocol and addressing) to all ONodes of a SSON, so that they can communicate with each other independent of their differences regarding the underlying protocol stacks and technologies. On top of the OSL, and using its services, there are application modules that implement the behavior of a MC, MS, or MP in regard to data handling. MCs act as data sinks and send the multimedia data to the end-point media applications, whereas MSs act as data sources and receive the multimedia data from the end-point applications.

Service-Specific Overlay Networks Management

In SMART an overlay service is formed from the integration of multiple components in the network. For instance, a specific set of overlay nodes and their resources in addition to a specific set of the required MediaPorts are needed to construct a single SSON. A certain overlay node could be part of many SSONs, and the network will run hundreds of SSONs at the same time. Thus, if left unmanaged, they will degrade the performance not only of each other, but also of the underlying network. Moreover, each SSON has to meet a certain QoS parameters defined by the user and the service provider and has to adapt itself to the changing condition of its environment. Therefore, the management of these services is not a trivial or an easy task, while current manual approaches to service management are costly and consume resources and IT professionals' time, which leads to increased customer dissatisfaction. Therefore, autonomic computing (AC) is proposed in the next section to address the problem of service management complexity.

AN AUTONOMIC SERVICE ARCHITECTURE

Given the multiple sources of the heterogeneity of networks, users, and applications, constructing and managing SSONs in large distributed networks is challenging. Media content usually requires adaptation before it is consumed by media clients. For example, video frames must be dropped to meet QoS constraints. Other examples are when a client with a Personal Digital Assistant (PDA) requires a scale-down for a video, or when content must be cached to be viewed by a mobile user. In addition to adaptation, new users may request to join or leave the overlay, a network node may fail, or a bottleneck may degrade the SSON's performance. Consequently, the overlay must be adapted to overcome these limitations and to satisfy the new requests. It is obvious that with a large number of overlays, the management task becomes harder to achieve using traditional methods. Therefore new solutions are needed to allow SPs to support the required services and to focus on enhancing these services rather than their management.

Moreover, IT professionals must reinforce the responsiveness and resiliency of service delivery by improving quality of service while reducing the total cost of their operating environments. Yet, information technology (IT) components over the past decades are so complex that they increase the challenges to effectively operate a stable environment. Overlay networks management complexity is turn increased by the huge number of users, terminals, and services. Although human intervention enhances the performance and capacity of the components, it drives up the overall costs—even as technology component costs continue to decline. Because of this increased management complexity, AC is a key solution for SSON management in heterogeneous and dynamic environments.

Introduction

A service delivered to a customer by a service provider (SP) is usually composed of different services. Some services are basic in the sense that they cannot be broken down further into component services, and they usually act on the underlying resources. Other services are composed of several basic services, each consisting of an allocation of resource amounts to perform a function. However, with the increasing demands for QoS, service delivery should be efficient, dynamic, and robust. Current manual approaches to service management are costly and consume resources and IT professionals' time, which leads to increased customer dissatisfaction; the advent of new devices and services has further increased the complexity. With a large number of overlays, the management task becomes harder to achieve using traditional methods. Therefore, new solutions are needed to allow SPs to support the required services and to focus on enhancing these services rather than their management. AC helps address this complexity by using technology to manage technology.

IBM proposed the concept of autonomic computing [14] to enable systems to manage themselves through the use of self-configuring, self-healing, self-optimizing, and self-protecting solutions. It is a holistic approach to computer systems design and management, aiming to shift the burden of support tasks, such as configuration and maintenance, from IT professionals to technology. Therefore, AC is a key solution for SSON management in heterogeneous and dynamic environments.

Establishing an SSON involves (1) resource discovery to discover network-side nodes that support the required media processing capabilities, (2) an optimization criterion to decide which nodes should be included in the overlay network, (3) configuring the selected overlay nodes, and (4) adapting the overlay to the changing network context, user, or service requirements, and joining and leaving nodes. In AC, each step must be redesigned to support autonomic functions. In other words, in autonomic overlays (AOs), each step imposes a set of minimum requirements. For example, the resource discovery scheme should be distributed and not rely on a central entity. It needs to be *dynamic* to cope with changing network conditions; *efficient* in terms of response time and message overhead; and *accurate* in terms of its success rate. The optimization step is mapped into a self-optimization scheme that selects resources based on an optimization criterion (such as delay, bandwidth, etc.) and should yield the cheapest overlay and/or an overlay with the least number of hops; an overlay that is load-balanced; a low-latency overlay network; and/or a high-bandwidth overlay network. The configuration of the selected overlay nodes in a given SSON is mapped into a self-configuration and self-adaptation. Self-configuring SSONs dynamically configure themselves on the fly. Thus they can adapt their overlay nodes immediately to the joining and leaving nodes and to the changes in the network environment. Self-adapting SSONs self-tune their constituent resources dynamically to provide uninterrupted service. Our goals are to automate overlay management in a dynamic manner that preserves the flexibility and benefits that overlays provide, to extend overlay nodes to become autonomic, to define the internode autonomic behavior between overlay nodes, and to define the global autonomic behavior between SSONs.

In the following section we present a novel management architecture for overlay networks. The architecture offers two main contributions. First, we introduce the concept of autonomic overlays (AOs), in which SSONs and their constituent overlay nodes are made autonomic and thus become able to self-manage. Second, autonomic entities are driven by policies that are generated dynamically from the context information of the user, network, and service providers. This ensures that the creation, optimization, adaptation, and termination of overlays are controlled by policies, and thus the behaviors of the overlays are tailored to their specific needs.

Autonomic Overlays

To tackle the complexity of overlay management, each SSON is managed by an SSON autonomic manager (SSON-AM) that dictates the service performance parameters. This ensures the self* functions of the service. In addition, overlay nodes are made autonomic to self-manage their internal behavior and their interactions with other overlay nodes. In order to ensure systemwide performance, System Autonomic Managers (SAM) manages the different SSON managers by providing them with high-level directives and goals. The following sections detail the different aspects of our architecture.

Architecture Overview

The components that make up our architecture are shown in Figure 2.3. The lowest layer contains the system resources that are needed for multimedia delivery

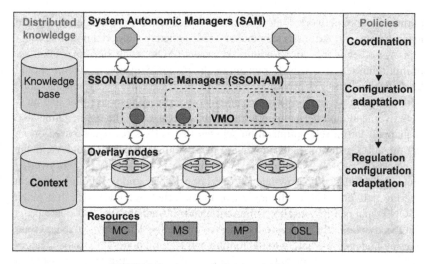

FIGURE 2.3   Autonomic Overlays Architecture.

sessions. In particular, the Overlay Support Layer (OSL) receives packets from the network, sends them to the network, and forwards them on to the overlay. Overlay nodes implement a sink (MediaClient, or MC), a source (MediaServer, or MS), or a MediaPort (MP) in any combination. MPs are special network-side components that provide valuable functions to media sessions; these functions include, but are not limited to, special routing capabilities, caching, and adaptation. These managed resources can be hardware or software and may have their own self-managing attributes.

The next layer contains the overlay nodes. Overlay nodes are physical Ambient Network nodes that have the necessary capabilities to become part of the SSON. They consist of a control plan and a user plan. The control plan is responsible for the creation, routing, adaptation, and termination of SSONs, while the user plan contains a set of managed resources. The self-management functions of overlay nodes are located in the control plan. The self-managing functions use the Ambient Manageability interfaces to access and control the managed resources. The rest of the layers automate the overlays' management in the system using their autonomic managers. SSON-AMs and SAMs may have one or more autonomic managers, for example, for self-configuring and self-optimizing. Each SSON is managed by an SSON-AM that is responsible for delivering the self-management functions to the SSON. The SAMs are responsible for delivering systemwide management functions; thus, they directly manage the SSON-AMs. The management interactions are expressed through policies at different levels. All of these components are backed up with a distributed knowledge. The following sections describe each component in detail.

Autonomic Elements

Overlay Nodes Autonomic Manager (ONAM)

Each overlay node contains a control loop similar to the IBM control loop [10], as shown in Figure 2.4. The autonomic manager (AM) collects the details it needs from its managed resources, analyzes those details to decide what actions need to change, generates the policies that reflect the required change, and enforces these policies at the correct resources. As shown in the figure, the ONAM consists of the following:

Monitoring Agents (MAs): These agents collect information from the overlay node resources, such as packet loss, delay jitter, and throughput. A MA also correlates the collected data according to the installed policies and reports any violation to the Analyze/Learning Agent (ALA). For example, an MA for a Caching MP collects information about the MP's available capacity, and whenever the available capacity reaches 10%, it reports to the ALA. Another example is the MA for a routing MP that relays data packets between overlay nodes: Its MA collects information about the throughput and reports to the ALA whenever the throughput reaches a high value. These collected data will be used to decide the correct actions that must be taken to keep the overlay node performance within its defined goals. The MAs interact with the Resource Interface Agents (RIAs) to monitor the overlay node resources availability and to collect data about the desired metrics. They also receive policies regarding the metrics that they should monitor as well as the frequency with which they report to the ALA.

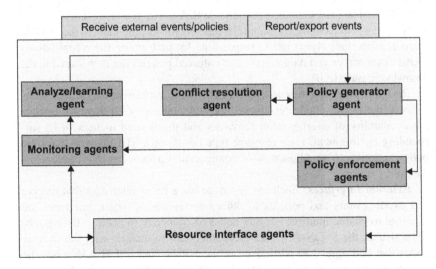

FIGURE 2.4 Autonomic Control Loop.

Analyze/Learning Agent (ALA): This agent observes the data received from the MAs and checks to see whether a certain policy with which its overlay node is associated is being met. It correlates the observed metrics with respect to the contexts and performs analysis based on the statistical information. In the event that one of policies is violated, it sends a change request to the Policy Generator (PG). This component is an objective of future work.

Policy Generator (PG): The difference between this control loop and IBM's control loop lies in the use of a PG instead of a Plan component. The Plan function—according to IBM [10]—is to select or create a procedure that reflects the desired change based on the received change request from the Analyze Agent. This is not sufficient in our case, where each overlay node receives high-level policies, and it is up to the overlay node to decide how to enforce these policies based on its available resources. Therefore, we envisioned a PG instead. The PG reacts to the change request in the same way as in the Plan component, although it also generates different types of policies in response to the received high-level policies. For example, based on the goal policies received by the overlay node, the policy generator generates the tuning polices and passes them to the MAs (more about this is presented in an earlier section). Upon generating new policies, the policy generator consults a Conflict Resolution Agent (CRA) that ensures the consistency of the new generated policies with those that already exist. Generally, we divide conflicts into two types: static and dynamic. In our model, a static conflict is a conflict that can be detected at the time a new policy is generated, while a dynamic conflict is one that occurs at runtime.

Policy Enforcement Agent (PEA): The PG generates suitable policies to correct the situation in response to a change request and passes these policies to the PEA. The PEA then uses the suitable RIA to enforce them. This includes mapping the actions into executable elements by forwarding them to the suitable Resource Interface Agent (RIA) responsible for performing the actual adjustments of resources and parameters. The enforced policies are then stored in the Knowledge Base (KB).

Resource Interface Agents (RIAs): The RIAs implement the desired interfaces to the overlay node resources. The MAs interact with them to monitor the availability of overlay node resources and the desired metrics in its surrounding environment. Each resource type has its own RIA that translates the policy actions into an adjustment of configuration parameters that implements the policy action.

External Interfaces: Each overlay node has a set of interfaces that receive and export events and policies to other overlay nodes. These interfaces are essential to enable multiple overlay nodes to cooperate to achieve their goals. In particular, the SSON-AM uses these interfaces to interact with the overlay nodes that had agreed to participate in the SSON. The SSON-AM sends the system policies to the overlay nodes through these interfaces, through which it also receives reports on their current status.

SSON Autonomic Managers (SSON-AM)

SSON-AMs implement the intelligent control loop in much the same way as ONAMs. They automate the task of creating, adapting, configuring, and terminating SSONs. They work directly with the ONAM through their management interfaces, and they perform different self-management functions, such as self-configuring, self-optimizing, and self-adapting. Therefore, they have different control loops. Typically, they perform the following tasks:

Self-configuration: SSON-AMs generate configuration policies in response to the received system policies. They use these policies to configure overlay nodes that are participating in a given SSON.

Self-optimization: During SSON construction, SSON-AMs discover the overlay nodes required to set up a routing path for the multimedia session. Therefore, they are responsible for optimizing the service path to meet the required QoS metrics induced from high-level policies as well as the context of the service.

Self-Adaptation: SSON-AMs monitor the QoS metrics for the multimedia session and keep adapting the service path to the changing conditions of the network, service, and user preferences. They also monitor the participating overlay nodes and find alternatives in case one of the overlay nodes is not conforming to the required performance metrics.

SSON-AMs receive goal policies from the SAMs to decide the types of actions that should be taken for their managed resources. A SSON-AM can manage one or more overlay nodes directly to achieve its goals. Therefore, the overlay nodes of a given SSON are viewed as its managed resources. In addition, they expose manageability interfaces to other autonomic managers, thus allowing SAMs to interact with them in much the same way that they interact with the ONAMs.

This is illustrated in Figure 2.5 where the lower part represents an SSON that consists of a Source (S), a Destination (D), and a MediaPort (MP). The SSON is managed by a SSON-AM. Since the SSON-AM can manage multiple SSONs, it has its own Knowledge Base (KB). It also contains a PG backed up with a Conflict Resolution Agent (CRA). The PG has access to the available context information that helps it achieve its goals. The upper part represents a SAM and its components. The SAM is able to manage one or more SSON-AMs. Therefore, it has its own KB and PG. The context information of the user, network, and service is assumed to be available to these autonomic managers as they can acquire it from the Context Functional Area in the Ambient Control Space [13].

System Autonomic Managers (SAM)

A single SSON-AM alone is only able to achieve self-management functions for the SSON that it manages. If a large number of SSONs in a given network with their autonomic managers is considered, it is observable that these SSONs are not really isolated. On the one hand, each overlay node can be part

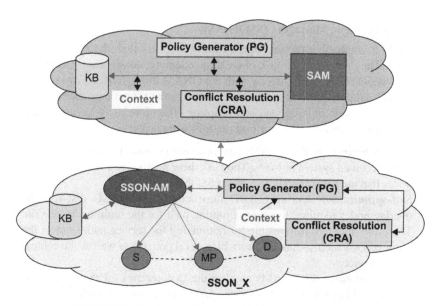

FIGURE 2.5 The Relation between an SSON, SSON-AM, and SAM.

of many SSONs if it offers more than one service or if it has enough resources to serve more than one session. On the other hand, the SSONs' service paths may overlap, resulting in two or more SSONs sharing the same physical or logical link. For example, consider two SSONs sharing the same routing MP with the same goal to maximize throughput. This will lead to a competition between autonomic managers that are expected to provide the best achievable performance. Therefore, and in order to achieve a systemwide autonomic behavior, the SSON-AMs need to coordinate their self-managing functions. Typically, this is achieved using SAMs. SAMs can manage one or more SSON-AMs. They pass the system high-level policies, such as load balancing policies, to the SSON-AMs. Moreover, whenever they find shared goals between two different SSON-AMs, they inform them to avoid conflicting actions. The involved autonomic managers then contact each other to coordinate their management actions before they are passed to their overlay nodes.

Sharing goals is not the only reason for the coordination step; SSONs sharing common links as well as SSONs that belong to the same policy domain (same service class, ISP, etc.) may also need to coordinate their management actions. Moreover, SSONs that share common nodes/links affect each other's performance, as they compete for the shared resources. This might result in a degraded performance as the competition will cause the control loop to be invoked frequently in an attempt to reach the desired performance goals. Also, all the SSONs in a given domain (ISP) are expected to achieve the domain-wide policies together. Coordination allows these policies to be dispatched and adapted to each SSON in a way that achieves the desired goals. Moreover, it also allows the sharing of control and information between different SSONs. A set of SSONs that are co-located in a given vicinity (such as an area, domain, AS, etc.) are usually equipped with independent route decisions based on their observations of their environment. Sharing this information will result in reduced overhead for each overlay to compute this information and will allow for adapting and generating policies to achieve better performance.

Distributed Knowledge

Each autonomic manager obtains and generates information. This information is stored in a shared Knowledge Base (KB) (see Figure 2.5). The shared knowledge contains data such as SSON topology, media-type descriptions, policies that are active, and goal policies received from higher level autonomic managers. The shared knowledge also contains the monitored metrics and their respective values. When coordination is needed, each autonomic manager can obtain two types of information from its peers. The first is related to the coordination actions, and the second to the common metrics in which each autonomic manager is interested. Therefore, knowledge evolves over time; the autonomic manager's functions add new knowledge as a result of executing their actions, and obsolete knowledge is deleted or stored in log files. Also, goal policies are passed from high-level autonomic managers to their managed autonomic managers. The context information of the network, users, and services is also used primarily to aid in generating suitable policies at each level of autonomic managers.

Policies

The use of policies offers an appropriately flexible, portable, and customizable management solution that allows network entities to be configured on the fly. Usually, network administrators define a set of rules to control the behavior of network entities. These rules can be translated into component-specific policies that are stored in a policy repository and can be retrieved and enforced as needed. Policies represent a suitable and an efficient means of managing overlays. However, the proposed architecture leverages the management task to the overlays and their logical elements, thus providing the directives on which an autonomic system can rely to meet its requirements. Policies in our autonomic architecture are generated dynamically, thereby achieving an automation level that requires no human interaction. In the following, we will highlight the different types of policies specific to autonomic overlays. These policy types are generated at different levels of the system.

Configuration policies: These are policies that can be used to specify the configuration of a component or a set of components. The SSON-AMs generate the configuration polices for the service path that meets the SSON's QoS requirements. The ONAMs generate the specific resource configuration policies that, when enforced, achieve the SSON QoS metrics. These autonomic managers employ the user, service, and network context to generate configuration policies.

Adaptation policies: These policies are those that can be used to adapt the SSON to changing conditions. They are generated in response to a trigger fired by a change in the user, service, or network context. SSON-AMs receive these triggers either from the SAMs or from the ONAMs, while the ONAMs receive these triggers either from the SSON-AMs or from their internal resources. Whenever a change that violates the installed policies occurs, an adaptation trigger is fired. The autonomic manager that first detects this change tries to solve the problem by generating the suitable adaptation policies; if it does not succeed, it informs the higher level autonomic manager.

Coordination policies: Coordination policies can be used to coordinate the actions of two or more SSON-AMs. They are generated by the SAMs to govern the behavior of SSON managers that have conflicting goals to avoid race conditions.

Regulation policies: These policies are generated by the overlay nodes themselves to control the MAs' behavior with respect to their goals. For example, an MA that measures throughput has a policy to report throughput <70%. Another regulation policy can be installed to replace this policy and report throughput <90%. The second regulation policy can be generated in response to an adaptation policy that requires throughput to be at least 90%. The MAs therefore are made more active to contribute to achieving the required tasks.

Figure 2.6 shows how these policies are related to our autonomic architecture. At the highest level, the SAMs define the set of system polices. These policies represent the systemwide goals and do not describe either the particular devices that will be used to achieve the system goals or the specific configurations for these devices. SAMs pass these policies to the SSON-AMs.

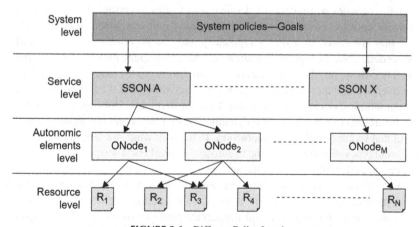

FIGURE 2.6 Different Policy Levels.

SSON-AMs refine the system policies and generate service-specific policies. They do so by adding further details to the system policies. These details are induced from the system policies as well as from the context information of the users, network, and service. At this level, the goals of the SSON under discussion, such as the permitted QoS metrics, are defined. These goals are still device-independent policies. The set of service polices is then passed to the ONAMs. These autonomic managers further refine the received policies and generate the overlay node polices and their respective resource-specific policies. Overlay node policies represent the goals that this overlay node is expected to achieve, while resource-specific policies represent the actual actions that the resources of the overlay node have to perform to achieve the overlay node goals. This separation of policies allows each autonomic element to focus on its goals and on how to achieve them using its current resources while contributing at the same time to the overall system performance. By de-coupling the functionality of adapting overlay node resources policies from the task of mapping system objectives and abstract users' requirements, the policy separation offers users and IT professionals the freedom to specify and dynamically change their requirements. The hierarchical policy model is used to facilitate the mapping of higher level system policies into overlay node objectives. Given sets of user, service and network context and constraints, as well as sets of possible actions to be taken, decisions for policy customizations are taken at runtime based on values obtained from MAs to best utilize the available overlay node resources.

In addition to generating policies from high-level goals, the policy generator located in each autonomic manager serves as a Policy Decision Point (PDP) for the low-level autonomic manager. For example, the SSON-AM serves as a PDP for the ONAM. Whenever an ONAM detects that one of the configuration policies has been violated, it tries to solve the problem locally. If it is unable to do so, it consults the SSON-AM to which the overly node is providing a service. The SSON-AM then tries to solve the problem by either relaxing the goals of the services or finding an alternative overlay node that is able to achieve the SSON's goals. The SSON-AM then informs the ONAM of its decision and may also consult its designated SAM to acquire decisions on situations that it cannot handle locally. The autonomic manager acting as a PDP decides which policies, if any configuration or adaptation policies have been violated, were most important and what actions to take. It uses information about the installed policies and the current context of the user, network, and service.

CONCLUSION

This chapter has presented a novel scheme for SSONs autonomic management. This work provides a complete integrated architecture for autonomic SSONs management; it illustrates the benefits of avoiding the complexity of existing service management systems. The road toward fully autonomic system architecture is still long; however, this chapter presents an autonomic overlay architecture that represents the basic building blocks needed by autonomic overlay systems.

The success of autonomic computing relies on a system's ability to manage themselves and to react to changing conditions. The proposed layered architecture for autonomic overlay provision enables autonomy and dynamic overlay construction through multilevel policies. The architecture components can self-assemble into an overall autonomic system—flexibility is crucial to the system. Therefore, individual overlay nodes should be able to self-organize to form diverse SSONs. This is possible through investigation of the different media types and QoS requirements for each media delivery session, which allows for the dynamic self-composition of the fundamental services needed by SSONs. This will lead to the ultimate dynamic self-management and will require the dynamic assignment of SSON-AMs and SAMs.

REFERENCES

[1] F. Hartung, S. Herborn, M. Kampmann, and S. Schmid, "Smart Multimedia Routing and Adaptation using Service Specific Overlay Networks in the Ambient Networks Framework," Proceedings 12. Wireless World Research Forum (WWRF) Meeting, November 2004, Toronto/Canada.

[2] G. Cortese, R. Fiutem, P. Cremonese, S. D'antonio, M. Esposito, S.P. Romano, and A. Diaconescu, "Cadenus: creation and deployment of end-user services in premium IP networks," Communications Magazine, IEEE, Vol. 41, No. 1, pp. 54–60, January 2003.

[3] K. Shami, D. Magoni, and P. Lorenz, "A Scalable Middleware for Creating and Managing Autonomous Overlays," Communication Systems Software and Middleware, 2007. COMSWARE 2007. 2nd International Conference on, Vol., No., pp. 1–8, 7–12 January 2007.

[4] K. Ragab, N.Y. Horikoshi, H. Kuriyama, and K. Mori, "Autonomous decentralized community communication for information dissemination," Internet Computing, IEEE, Vol. 8, No. 3, pp. 29–36, May–June 2004.

[5] K. Ragab, Y. Horikoshi, H. Kuriyama, and K. Mori, "Multi-layer autonomous community overlay network for enhancing communication delay," Computers and Communications, 2004. Proceedings. ISCC 2004. Ninth International Symposium on, Vol. 2, No., pp. 987–992, 28 June–1 July 2004.

[6] S. Ratnasamy, M. Handley, R. Karp, and S. Shenker, "Topologically-aware overlay construction and server selection," INFOCOM 2002. Twenty-First Annual Joint Conference of the IEEE Computer and Communications Societies. Proceedings. IEEE, Vol. 3, No., pp. 1190–1199, 2002.

[7] C.-J. Lin, Y.-T. Chang, S.-C. Tsai, and C.-F. Chou, "Distributed Social-based Overlay Adaptation for Unstructured P2P Networks," IEEE Global Internet Symposium, 2007 , Vol., No., pp. 1–6, 11 May 2007.

[8] J.A. Pouwelse, P. Garbacki, J.W.A. Bakker, J. Yang, A. Iosup, D. Epema, M. Reinders, M.R. van Steen, and H.J. Sips, "Tribler: A social-based based peer to peer system," in 5th Int'l Workshop on Peer-to-Peer Systems (IPTPS), February 2006.

[9] P. Androutsos, D. Androutsos, and A.N. Venetsanopoulos, "Small world distributed access of multimedia data: an indexing system that mimics social acquaintance networks," Signal Processing Magazine, IEEE, Vol. 23, No. 2, pp. 142–153, March 2006.

[10]   IBM Corporation, "An architectural blueprint for autonomic computing," White Paper, June 2006.

[11]   N. Niebert, A. Schieder, H. Abramowicz, G. Malmgren, J. Sachs, U. Horn, C. Prehofer, and H. Karl, "Ambient networks: an architecture for communication networks beyond 3G," *Wireless Communications, IEEE*, Vol. 11, No. 2, pp. 14–22, April 2004.

[12]   N. Niebert, M. Prytz, A. Schieder, N. Papadoglou, L. Eggert, F. Pittmann, and C. Prehofer, "Ambient networks: a framework for future wireless internetworking," Vehicular Technology Conference, 2005. VTC 2005-Spring. 2005 IEEE 61st, Vol. 5, No., pp. 2969–2973, 30 May–1 June 2005.

[13]   W.T. Ooi, R.V. Renesse, and B. Smith, "The design and implementation of programmable media gateways," in: Proc. NOSSDAV'00, Chapel Hill, NC, June 2000.

[14]   J.O. Kephart and D.M. Chess, "The vision of autonomic computing," *Computer*, Vol. 36, No. 1, pp. 41–50, January 2003.

Chapter 3

ANA: Autonomic Network Architecture

Prométhée Spathis and Marco Bicudo
UPMC Univ Paris 6, UMR 7606, LIP6/CNRS, F-75005, Paris, France
Emails: promethee.spathis@lip6.fr, marco.bicudo@lip6.fr

INTRODUCTION

The Autonomic Network Architecture (ANA)[1] integrated project is co-sponsored by the European Commission under the Information Society Technology (IST) priority on Future and Emerging Technologies (FET) under the 6th Framework Programme.

The ANA project aims at exploring novel ways of organizing and using networks beyond legacy Internet technology. The ultimate goal is to design and develop a novel network architecture that can demonstrate the feasibility and properties of autonomic networking. The project addresses the self-* features of autonomic networking such as self-configuration, self-optimisation, self-monitoring, self-management, self-repair, and self-protection. This is expected to be especially challenging in a mobile context where new resources become available dynamically, administrative domains change frequently, and the economic models may vary according to the evolution of technologies and services.

Designing a network architecture implies identifying and specifying the design principles of the system being developed. The architecture defines the atomic functions and entities that compose the network and specifies the interactions which occur between these various building blocks. This chapter describes the reference model of the ANA autonomic network architecture. It includes a definition of the basic abstractions and concepts of the ANA architecture and the communication paradigms. The chapter also includes the abstract interfaces that an actual implementation of a network node must provide in order to support the ANA architectural concepts, and defines the information flow model for the relevant control and management information of an autonomic network.

---

[1] http://www.ana-project.org

Autonomic Network Management Principles. DOI: 10.1016/B978-0-12-382190-4-4.00003-6

Motivation and Aims

The overall objective is to develop a novel network architecture and to populate it with the functionality needed to demonstrate the feasibility of autonomic
networking. In order to avoid the same shortcomings of network architectures
developed in the past, the guiding principles behind the architectural work in
ANA are achieving maximum *flexibility* and providing support for *functional
scaling* by design.

The former indicates that the project members do not envision a "one-
size-fits-all" network architecture for the various different types of networking
scenarios (e.g., Internet, Sensor Networks, Mobile Ad hoc Networks, Peer-
to-peer Networks). Instead, the aim of ANA is to provide an architectural
framework that enables the co-existence and interworking between different
"networking style".

Functional scaling means that a network is able to extend both horizontally
(more functionality) as well as vertically (different ways of integrating abun-
dant functionality). New networking functions must be integrated into the "core
machinery" of the system, and not on top as "patches'"; otherwise we do not
have scaling of functionality, only function accumulation on top of an ossified
network core. In the context of ANA, the ability of functional scaling is considered vital, as this provides the basis to allow dynamic integration of new,
autonomic functionalities, as they are developed.

The Internet and the Open Systems Interconnection (OSI) reference model
are examples of a functionally non-scalable approach: both are based on the
premise of factoring out functionality with a pre-defined and static model. The
"canonical set" of protocols collected in the Internet-Suite has done a good job
so far; however the limitations of this one-size-fits-all approach have become
visible. This is not surprising, since the networking layer of the Internet has
not aimed at functional scaling by design: it has evolved through a 'patchwork'
style. Additions were made in a stealth-like way and have not dared to change
the core of IP networking. The main examples are the introduction of the hidden
routing hierarchy with AS, CIDR or MPLS, as well as other less successful
initiatives such as RSVP, multicast, mobile IP and MANET.

As a result, the objective of the ANA project is to provide an architectural framework a "meta architecture" that allows the accommodation of and
interworking between the full range of networks, ranging from small scale Personal Area Networks, through (Mobile) Ad hoc Networks and special purpose
networks such as Sensor Networks, to global scale networks, in particular the
Internet.

An important step towards such a "meta architecture" is to design a framework that is able to host different types of networks. As a result, ANA encompasses the concept of *compartments* as a key abstraction that allows co-existence
and interworking of different types of network through a minimum generic
interface.

Furthermore, the operation of different types of compartment must be analysed in order to identify the fundamental building blocks of the abstraction. The decomposition of compartments into the core functions helps to understand how the necessary flexibility and functional scalability to provide *autonomicity* can be achieved through compartments.

Scope and Structure

This chapter focus on the overall architectural aspects of ANA. As such, they define the basic abstractions and building blocks of ANA and present their basic operation and interactions. Therefore, the chapter does not capture the detailed results of other areas of work in the project, such as routing, service discovery, functional composition, monitoring, etc.

This chapter is organized as follows. Section 2 describes the core abstractions and basic communication paradigms of the ANA network architecture. Section 3 introduces the ANA Compartment API which is used to wrap different "network types" with a generic interface. Section 4 captures the prototyping activity in ANA by describing the implementation of a routing scheme that offers a global network connectivity over heterogeneous environments. Finally, Section 5 concludes the document.

CORE ARCHITECTURAL ABSTRACTIONS

This section introduces the basic abstractions of ANA: the functional block (FB), the information dispatch point (IDP), the compartment, and the information channel (IC).

Basic abstractions

Functional Block (FB)

In ANA, any protocol entity generating, consuming, processing and forwarding information is abstracted as a functional block (FB). For example, an IP stack, a TCP module, but also an encryption function or a network monitoring module can be abstracted as a functional block. In contrast to OSI's protocol entities [1], functional blocks are not restricted to being abstractions of network protocols and the functionality they provide can range from a full monolithic network stack down to a small entity that computes checksums.

The fact that a FB can represent the whole range from an individual processing function to a whole network stack makes this abstraction very useful. It permits the capture and abstraction of many different flavours of communication entities and types, and enables them to interact with other entities inside ANA under a generic form. As shown in Fig. 3.1, a functional block (FB) is represented by a grey-shaded square.

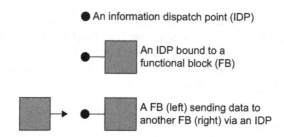

FIGURE 3.1   Functional Block and Information Dispatch Points.

Information Dispatch Point (IDP)

The fundamental concept introduced by ANA and around which the core architecture machinery is designed is the information dispatch point (IDP). IDPs are inspired by network indirection mechanisms [2], and they are somehow similar to file descriptors in Unix systems. Basically inside an ANA node, a functional block (FB) is always accessed via one or multiple IDPs attached to it: in other words, all interactions are carried out via an indirection level built into the network architecture. In the same way as Unix systems enforce access to all system components (e.g., devices, sockets, files) via file descriptors, ANA mandates that all functional blocks are accessed via IDPs. However unlike file descriptors and sockets, the binding of an IDP (i.e. the FB to which it is attached) is dynamic and can change over time (e.g., if a "network stack" is reconfigured). The bindings between IDPs and FBs are stored in the core forwarding table of the ANA Node where each IDP is identified by a node-local label (e.g., a 32-bit unsigned integer value). As detailed later, each IDP also maintains some dynamic information such as the FB to which it is attached, state to be passed to the FB, a lifetime counter, and some status information.

The advantage of IDPs is two-fold: first, they act as generic communication pivots between the various functional blocks running inside an ANA node and, second, they provide the required flexibility allowing for the re-organization of communication paths. For packet forwarding, IDPs permit the implementation of forwarding tables which are fully decoupled from addresses and names: i.e., in ANA the next hop entity (local or remote) is always identified by an IDP. This permits the easy addition and use of new networking technology and protocols as long as they export their communication services as IDPs.

As shown in Fig. 3.1, an information dispatch point (IDP) is represented by a black dot. When an IDP is bound to a functional block (FB), the binding is represented by a line connecting the IDP to the FB. Essentially, this means that any data sent to, or any action applied to an IDP is really received or captured by the functional block it is attached to. ANA strictly enforces this level of indirection: that is, it is not possible to interact directly with a functional block. All interactions are always handled via an IDP. In Fig. 3.1, the action of sending to or acting on an IDP is drawn as an arrow.

Compartment and Information Channel (IC)

Background and Motivation

Recent research has led to the conclusion that todays networks have to deal with a growing diversity of commercial, social and governmental interests that have led to increasingly conflicting requirements among the competing stakeholders. A pragmatic way to resolve those conflicts is to divide the network into different realms [3] that isolate the conflicting or competing interests.

In order to abstract the many notions of network realms and domains, ANA introduces the key concept of compartment. With respect to communication networks, the concept of compartment is similar to the notion of context [4] where "a context describes a region of the network that is homogeneous in some regard with respect to addresses, packet formats, transport protocols, naming services, etc."

The compartment abstraction, as one of the fundamental concepts of ANA, enables ANA to "encapsulate" (or "wrap") todays networks into a *generic* abstraction. Compartments indeed offer a set of generic "wrapper primitives" that allows accommodation of both legacy network technologies and new communication paradigms within ANA in a generic manner. Since support for backwards compatibility and inter-working with already deployed networks is a key requirement for any new network architecture, some architectural compromises (e.g., number of primitives and arguments of the API) are typically made to achieve this. However, as the compartment abstraction permits hiding the internals and specifics of individual network technologies, this concept enables interworking among heterogeneous compartments in a generic manner.

ANA Compartment: Definition

Prior to defining compartments and their properties, it is important to note that the notion of compartment does not necessarily involve multiple nodes: a compartment can indeed involve a single node only, for example to provide node-local communications between software entities. Hence in the rest of this document, we refer to *network compartment* when multiple nodes are involved, while we will refer to the *node compartment* to describe a specific construct of ANA allowing access to node-local software entities via the set of compartment generic primitives.

In ANA, compartments implement the operational rules and administrative policies for a given communication context. The boundaries of a communication context, and hence the compartment boundaries, are based on technological and/or administrative boundaries. For example, compartment boundaries can be defined by a certain type of network technology (e.g., a specific wireless access network) or based on a particular protocol and/or addressing space (e.g., an IPv4 or and IPv6 network), but also based on a policy domain (e.g., a national health network that requires a highly secure boundary). ANA anticipates that many compartments co-exist and that compartments are able to interwork on various levels.

The complexity and details of the internal operation are left to each compartment. That is, compartments are free to choose internally the types of addressing, naming, routing, networking mechanisms, protocols, packet formats, etc. For example in ANA, typical network compartments are: an Ethernet segment, the public IPv4 Internet, a private IPv4 subnet, the DNS, peer-to-peer systems like Skype, and distributed web caching networks like Akamai.

Typically, the communicating entities inside a compartment are represented in the ANA architecture through functional blocks (FBs). They implement the functionality that is required to communicate within the compartment. As such, the FBs can also be regarded as the processing elements or functions hosted by an ANA node that constitute the "compartment stack".

Information Channel (IC)

Typically in each network compartment, a distributed set of functional blocks collaborates in order to provide communication services to other compartments and applications. Whatever the nature of the communication service (unicast, multicast, datagram, reliable, etc.), it is abstracted by an *information channel* (IC) which (as anything else in ANA) is accessed via an IDP. While such an IDP is really bound to a functional block belonging to the network compartment, the abstraction provided is an information channel to some remote entity or entities. Note that the end-point of an information channel is not necessarily a network node: it can be a distributed set of nodes or servers, a software module, a routing system, etc. This rather abstract definition allows us to capture and abstract many different flavours of communication forms.

Information Channels (ICs) can be of either a physical or logical nature. Examples of physical ICs are a cable, a radio medium, or memory. A logical (or virtual) IC can represent a chain of packet processing elements and further ICs. The IC abstraction captures various types of communication channels, ranging from point-to-point links or connections, over multicast or broadcast trees to special types of channels such as anycast or concast trees.

The following example (see Fig. 3.2) shows how FBs, IDPs and ICs interact in order to enable communication within a compartment. The FBs in this example perform some type of processing (e.g., add protocol header, perform rate control, etc.) on the data, before it is transmitted to an entity in another ANA Node via the IC u → v.

The IC u → v in this example represents the service that is provided by some underlying system (e.g., the operating system providing a link and physical layer functionality). Since this IC is not provided by the network compartment considered in this example, it is shown in the imported view of the compartment. This IC can also represent the service provided by another, underlying compartment.

The IC s → t shows the "communication service" created by the network compartment as it would appear to an external entity using it. This IC hence

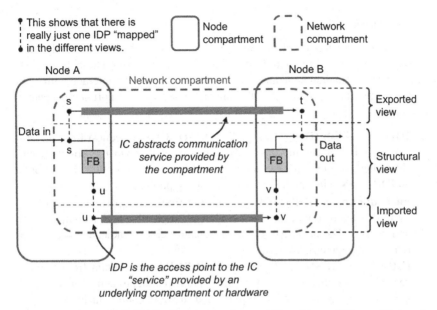

FIGURE 3.2 Different Views of a Compartment: Imported View, Structural View and Exported View.

appears in the exported view of the compartment: typically, the ICs exported by a compartment can be imported by other compartments.

Finally, the structural view of the figure shows the internal processing elements (i.e. FBs) that execute the protocols and functions that form the "compartment stack". Note that for simplicity the figure shows a single FB but in practice there could be multiple interconnected FBs providing the "compartment stack".

THE COMPARTMENT API

As described previously, network compartments in ANA are free to *internally* use their choice of addressing, naming, routing mechanisms, protocols, packet formats, etc. The objective is that ANA fosters variability and evolution by not imposing any restriction on the internal network operation of compartments.

However, to foster and guarantee all possible and unforeseen interactions within ANA, all compartments must support a generic *compartment API* which provides the "glue" that permits the building of complex network stacks and packet processing paths. With respect to application development, a key requirement is that the API supports generic function prototypes and arguments such that applications can use the same programming primitives when communicating to *any compartment*. In other words, a long-term objective of ANA is that developers can write "compartment-agnostic" applications that do not need to be modified to support new network compartments.

Basic Primitives

The ANA compartment API currently offers six fundamental primitives detailed below with some simplified C-style function prototypes. These primitives constitute the fundamental and generic interface used to interact with network compartments. In ANA, all compartments must support these generic primitives.

- **IDP$_s$ = publish (IDP$_c$, CONTEXT, SERVICE)**: to request from a compartment that a certain *service* becomes reachable via the compartment (identified by IDPc) in a certain *context*. When successful, `publish` returns an IDP (IDPs) via which the service has been published.
- **int = unpublish (IDP$_c$, IDP$_s$)**: to request from a compartment (identified by IDPc) that a certain *service* is no longer reachable via this compartment and *context*. The primitive returns an integer value to indicate a successful request or some error condition.
- **IDP$_r$ = resolve (IDP$_c$, CONTEXT, SERVICE)**: to obtain from a compartment (identified by IDP$_c$) an information channel to communicate with a certain *service* that has been published in some *context* inside the compartment. When successful, `resolve` returns an IDP (IDP$_r$) via which the remote service can be reached.
- **int = release (IDP$_c$, IDP$_r$)**: to release a previously resolved information channel identified by the IDP$_r$. The primitive returns an integer value to indicate a successful request or some error condition.
- **info = lookup (IDP$_c$, CONTEXT, SERVICE)**: to obtain further information from a compartment (identified by IDP$_c$) about a certain *service* that has been published in some *context* inside the compartment. When successful, `lookup` returns some information (to be detailed later) about the service being looked up.
- **int = send (IDP$_r$, data)**: to send data to a *service* which has been previously resolved into the IDP$_r$. The function returns an integer value to indicate a successful request or some error condition.

The two fundamental notions of CONTEXT and SERVICE have been introduced in order to be able to explicitly specify *what* (service) and *where* (context) "inside a compartment" a certain resource must be published or resolved. In other words, the context typically provides a hint to the compartment how to resolve the requested service while the service describes a resource inside the context. These notions are further detailed in the next section.

The Context and Service Arguments

The intuitive "design philosophy" behind the ANA API is that, for any kind of network applications and protocols, a user typically wants to find or reach something (the SERVICE) with respect to some scope (the CONTEXT) inside a network compartment. In order words, the SERVICE is what the user is looking for (or wants to communicate with, download, etc.), and the CONTEXT defines where and how (in which scope) the network operation (e.g., resolve) must be done. Based on our current experience with the API, we believe that it is possible to fit most if not all networking applications and protocols into the CONTEXT/SERVICE communication model.

IMPLEMENTATION OF A FUNCTIONAL BLOCK FOR INTER-COMPARTMENT CONNECTIVITY

Development Process

Prototyping in ANA is a key research activity which is fully part of the design process as it is used to test, validate, and refine the design of ANA over the course of the project. The ANA project was actually organised around two pro-totyping cycles. The objective of the first cycle was to develop a first prototype of ANA after the first half of the project. The result of this phase was the first public release of the ANA software in July 2008. This release contains the ANA core together with some functional blocks and an application that demonstrates that the ANA framework can be used. The second phase took place until the end of the project and focuses on the development of new functional blocks as well as on the refinement and the revision of the architecture based on the experience gained during the implementation of the new functional blocks. Our strategy was to embed the prototype implementation into the main research activities as a key for providing a viable (i.e. experimentally validated) architecture which could "grow" and mature over the course of the project. This contrasts with other projects where the implementation is often a final activity after which any architectural refinement is no longer possible.

The functional blocks that have been developed can be classified as follow. A first set of FBs are related to "data transmission", such as forwarding data, routing or interconnecting different network compartments. The second set focus on monitoring FBs that provide input for an autonomic configuration. Having these two major components, we can run the applications developed over the ANA framework, such as an address-agnostic chat and a video streaming service.

In the next section, we describe CCR, the functional block that permits inter-compartment routing based on content rather than hosts location.

Content Centric Routing

CCR is the functional block that offers a global network connectivity over heterogeneous environments. CCR's internal routing protocols are designed to be robust to a wide range of variance in network operations by adapting not only in the face of the machine failures and network partitions, but also in the presence of the active network attacks, ubiquitous mobility. CCR also offers a global network connectivity over a heterogeneous environment resulting from the composition of wildly heterogeneous types of networks.

In CCR protocol, clients issue keywords-based queries expressing their interest for some published content. Publications refer to the description of the content and some of its attributes. The network is composed of an overlay of Mediation Routers (MRs) that are responsible for matching client queries to stored publications and delivering the matching set to interested clients as well as disseminating publications. To do so, MRs implement routing tables called content dispatching tables (CDT) where they maintain states relative to the information conveyed by the exchanged messages (i.e. queries, publications, responses).

CCR's routing protocol is based on two main processes: filtering and dissemination. Filtering is the process that matches client queries against content publications. Filtering allows the router to select the best next hops to which the query has to be forwarded in order to reach the best matching publications. Dissemination is the process through which publications are forwarded towards MRs hosting. After these two processes are executed by the infrastructure, a delivery process is simply executed to send back the matching content publications to the interested users.

The next section describes CCR as implemented using the ANA framework. Following, we show which implementation tools have been used and all the details and the development of the brick that make the functional block content centric routing work.

CCR FB Modular Decomposition

CCR's component bricks

CCR's protocol is decomposed in a way that enables the client process to send CCR commands, to leave the FB do its own work, and, when the request is completely processed, to receive the response desired asynchronously. This is possible due to its bricks special design, which uses the MINMEX and Playground in a modular and flexible way. The protocol is a functional block, composed by smaller bricks that execute few and specific tasks. When put all together, those bricks work in harmony to perform the FB main purpose. The communication between the bricks as well as the nodes is encoded using XRP messages. As shown in Fig. 3.3, the CCR FB is decomposed in three main bricks: a CCR Service brick, a CCR Engine brick, and a CCR Dissemination brick. Each brick has specific features. The Service brick deals with all clients requests and manages network messages related to content request and content response. The Engine brick concentrates the CDT, all the information shared between Service and Dissemination bricks, as well as duplicated data structures of Service brick as tips to the routing algorithm, and executes the algorithm itself. The Dissemination brick is responsible for the content publishing and the keyword's dissemination, as the name stands for, and for the update of data structures and of the CDT in the Engine brick, which is crucial for a good functioning of the route decision algorithm.

Request

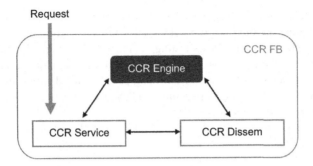

FIGURE 3.3   Decomposition of the CCR Functional Bloc.

Bricks interaction

As shown in Fig. 3.3, the Service brick receives requests from the clients. If the request is a content publish, the Service brick will update its local content table, which stores all pairs of keywords and contents locally published, and send a publish command to the Dissemination brick. The Dissemination brick will analyze the keywords; send an update message to Engine and then initiate the propagation over the network nodes the last published content.

If the Service brick receives a content request from a client, it consults the Engine brick about the next hop's IDP. The Engine brick in turn internally decomposes the keywords set into atomic sub-requests that contain only one keyword. The Mediation Routers process the sub-requests separately in order to make the routing process faster and simpler. The Engine brick chooses the best CDT entries for the next hop considering each atomic sub-request. As a result, the most pertinent and optimized IDP is returned back to the Service brick, which forwards the request or executes the corresponding action.

It is worth noticing that the Engine brick has also information about local published content, and it is considered during the decision making. It means that the Engine's response may actually inform the Service brick it has the matching content locally and, instead of forwarding the request, it should send back the content to the source node.

When a client request is received by the Service brick, it is processed as explained previously and the network search process is initiated. This process consists in routing the request through the network nodes and eventually hitting one node that recognizes the content's description. The data is then returned to the source node, more specifically, to the Service brick of the source node, using the reverse path that was constructed based in the req-id, which is an unique identifier. At that time, the Service brick will process the response, strip the content, and send it back to the client. This search protocol is done in a hop-by-hop fashion and is totally independent from the under layer network protocol, since we deal only with IDPs in CDT. It allows a more transparent and feasible communication among any sort of sub-networks that do not use necessarily IP addressing.

The benefits of the decomposition of CCR functionalities into separate functional blocks are obvious. It provides a flexibility of improving, testing and deploying each FB. It also allows testing the impact of each block on the other ones easily. But there are also some disadvantages of the existing decomposition: it increases the overhead of each sub-process called inside the routing process, the duplication of data structures and the use of messages, which could be avoided in one monolithic approach.

The next sections describe the internal implementation of the bricks and the operations of CCR.

Implementational Aspects

In this section, we present the specific tasks of each brick composing the CCR functional block. Then we describe different CCR communication scenarios with Ethernet or IP network.

CCR bricks

CCR Service brick

The Service brick publishes two or more IDPs in the IDT, depending on the number of Network Compartments. One IDP is supposed to receive client's specific messages. The others are related to the network's FBs found, which could be more than on in the case of an Ethernet Network Compartment.

CCR Engine brick

The Engine brick publishes only one IDP in the IDT. This IDP is supposed to receive all the updates and request messages from CCR Service and CCR Dissemination bricks.

CCR Dissemination brick

Dissemination brick publishes two or more IDPs in the IDT, depending on the number of Network Compartments. One IDP is supposed to receive Service brick's specific messages. The others are related to the network's FBs found, which could be more than on in the case of an Ethernet Network Compartment.

CCR operations

This section describes 5 use cases, namely: Retrieving a Content, Publishing/Unpublishing a Content, Treating a Publication, Treating a Content Request, and Treating a Content Response.

Retrieving a Content

To retrieve content, the client must first resolve the CCR FB IDP, which is in fact the Service brick IDP. After obtaining the IDP, the client then sends the Content Request command to the CCR FB in order to obtain the content desired. The

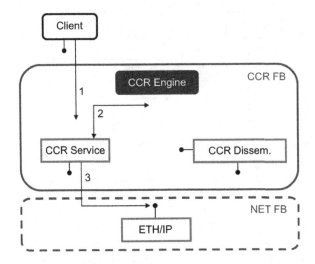

FIGURE 3.4   CCR Bricks Interactions During a Client Content Request.

CCR FB receives its command and triggers a sequence of procedures locally and subsequently some others remotely, as presented in Fig. 3.4.

1. Service brick receives client request
2. Service brick consults Engine brick about what is should do.
3. Service brick sends the request to the IDP informed by Engine brick.

Publishing/Unpublishing a Content

To publish content, the client must first resolve the CCR FB IDP, which is in fact the Service brick IDP. After obtaining the IDP, the client then sends the Content Publish command to the CCR FB in order to start the publication process of the desired content. The CCR FB receives its command and triggers a sequence of procedures locally and subsequently some others remotely, as presented in Fig. 3.5.

1. Service brick receives client's publish request.
2. Service brick notifies Dissemination brick about the content's keywords.
3. Dissemination brick updates Engine brick entries.
4. Dissemination brick initiates the content's keywords distribution over network, through network specific messages.

Treating a Content Publication/Unpublication

After receiving a client's request to publish, a CCR command is sent to the CCR neighbors, as presented previously. A network node that has a CCR FB enabled is supposed to receive CCR packets, treat it properly, and then forwarding the request it carries, if it is the case as presented in Fig. 3.6. The most important

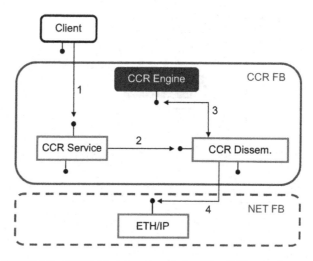

FIGURE 3.5   CCR Bricks Interactions During Client Publication Request.

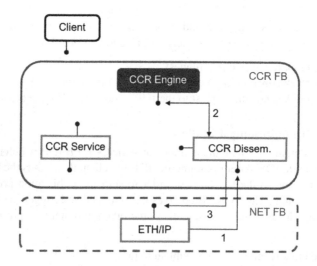

FIGURE 3.6   CCR Bricks Interactions During Network Publication Request.

task during the treatment of the CCR command is to keep up-to-date the Engine databases. This is actually the only task assigned to the Dissemination brick.

1. Dissemination brick receives a network publish packet.
2. Dissemination brick consults Engine brick about keys published.
3. Dissemination brick resends the publish packet based on the Engine response.

Treating a Content Request

After receiving a client's request to publish, a CCR command is sent to the CCR neighbors, as presented previously. A network node that has a CCR FB enabled is supposed to receive CCR packets, treat it properly, and then forwarding the request it carries, if it is the case. The most important task during the treatment of the CCR command is to keep up-to-date the Engine databases. This is actually the only task assigned to the Dissemination brick as shown Fig. 3.7.

1. Service brick receives network request.
2. Service brick extracts the keywords from the request, and submits them to Engine brick.
3. Based on its databases and the duplicated Service databases it decides the best action. Service brick:
   (a) Forwards the request.
   (b) Send a response to the request.

Treating a Content Response

This message is the response of a network node in order to answer a specific content request sent to the CCR network. It is received from the NC's IDP, as explained in the previous section. It triggers a sequence of messages as presented in Fig. 3.8.

1. Service brick receives a content response from network.
2. Service brick consults Engine brick updates Engines metrics.

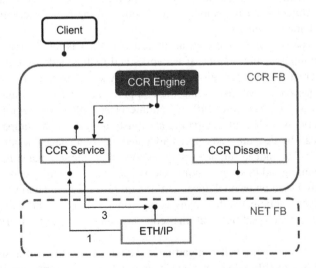

FIGURE 3.7   CCR Bricks Interactions During Network Content Request.

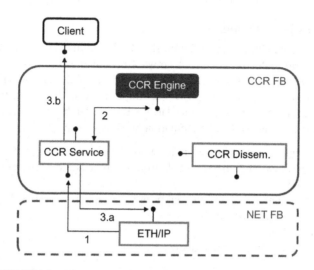

FIGURE 3.8   CCR Bricks Interactions During Network Content Response.

3. Based in its reverse path soft-states tables, Service brick:
   (a) Forwards the response to the next hop in the reverse-path.
   (b) Sends it back to the client, if it is a local request.

CONCLUSIONS

The main intention of this chapter is to describe the reference model of the ANA network architecture. This includes a description of the basic abstractions and communication paradigms.

At the coarsest level, the core abstraction of ANA is the "Compartment" which encompasses the notions of networks and realms and is the basic unit for the interaction and federation of networks. A key feature is that compartments reduce complexity by hiding the compartment internal communication complexity to the "outside world", which interfaces with a given compartment via a generic API that all compartments must support. With compartments, networks can be decomposed into smaller and easier manageable units: each compartment has indeed full autonomy on how to handle internal communication (i.e. naming, addressing, routing, etc.). The boundaries of a compartment are typically based on technological and/or administrative boundaries or policy domains. ANA anticipates that many compartments co-exist and will provide the necessary functionalities such that compartments are able to interwork on various levels.

At a more detailed level, compartments contain three kinds of abstraction: Functional Blocks (FBs), Information Channels (ICs), and Information Dispatch Points (IDPs). These abstractions are used to model the "internals" of a compartment and the service the compartment offers. The degree of details at which a compartment exposes its internal abstractions is specific to each compartment and ranges from full visibility to total opaqueness.

REFERENCES

[1] H. Zimmermann, Osi reference model - the iso model of architecture for open systems interconnection, IEEE Transactions on Communications 28 (4) (1980) 425–432.

[2] I. Stoica, D. Atkins, S. Zhuang, S. Shenker, S. Surana, Internet indirection infrastructure, in: Proc. of ACM Sigcomm'02, Pittsburg, USA, 2002.

[3] D. Clark, R. Braden, A. Falk, V. Pingali, Fara: Reorganizing the addressing architecture, in: Proc. of ACM SIGCOMM Workshop on Future Directions in Network Architecture (FDNA), Karlsruhe, Germany, 2003, pp. 313–321.

[4] R. M. T. R. J. Crowcroft, S. Hand, A. Warfield, Plutarch: an argument for network pluralism, in: Proc. of the ACM Workshop on Future Directions in Network Architecture, Karlsruhe, Germany, 2003.

Chapter 4

A Utility-Based Autonomic Architecture to Support QoE Quantification in IP Networks

Hajer Derbel, Nazim Agoulmine, Elyes Lehtihet, and Mikaël Salaün

ABSTRACT

The increasing scale, technology advances, and services of today's networks have dramatically complicated their management in such a way that in the near future human administrators will not be able to achieve efficient management. To control this complexity, a promising approach aiming to create self-managed networks has been introduced. This approach, called autonomic computing, aims to design network equipment able to self-adapt their configuration and to self-optimize their performance depending on their situation, in order to fulfill high-level objectives defined by human operators.

In this chapter, we present our Autonomic NEtwork Management Architecture (ANEMA) that implements several policy forms to achieve autonomic behaviors in the network. ANEMA captures the high-level objectives of the human administrators and users and expresses them in terms of *utility function* policies. The *goal* policies describe the high-level management directives needed to guide the network to achieve the previous utility functions. The *behavioral* policies capture behaviors that should be followed by network equipment to react regarding their context changes by considering the described goal policies.

In order to implement this utility-based analytical approach, we propose a set of extensions to the Distributed Management Task Force (DMTF) Common Information Model. These extensions represent the metrics used to calculate the Quality of Experience (QoE) as perceived by the user as well as the processes that implement it. This QoE will help the service provider to maximize the client's satisfaction while guaranteeing a good return on investment. In order to highlight the benefit of some aspect of ANEMA applied in case of multiservice IP network management, a testbed has been implemented and simulation scenarios have been achieved.

Keywords: Autonomic network management, Utility function policy, Goal policy, Behavioral Policy, Management context, QoS/QoE, DMTF CIM, Multiservice IP Networks.

INTRODUCTION

During the last decade, we have witnessed an enormous increase in the deployment and exploitation of IP networks. Consequently, the management of these networks has become a real challenge that is beyond the capacities of current humancentric network management approaches [1]. To control this increasing complexity and to make future operators able to manage their network infrastructure efficiently and reliably, it became urgent to investigate new approaches with decentralized and autonomic control. Autonomic computing proposed by IBM [2] [3] seeks to achieve this objective by allowing systems to self-manage their behavior in accordance with high-level objectives specified by human administrators [2]. The vast majority of IBM's work focuses on information technology issues, such as storage, database query performance, systems management, distributed computing, operations research, software development, artificial intelligence, and control theory [4]. In this chapter, we focus on IP networks and mechanisms that will allow them to exhibit autonomic management behaviors.

To make networks autonomic, it is important to identify the high-level requirements of the administrators. These requirements can be specified as utility functions that illustrate, according to Kephart [5], the practical and general way of representing them. At the next stage, it is necessary to map these utility functions to specific quality parameters and evaluate them based on information collected from the network. For that task, management architecture can employ some analytical optimization models. The specification of the constraints related to these functions and the way that they can be achieved can be guided by high-level management strategies. These strategies permit defining a second form of autonomic network management rules, called goal policies. These policies can be defined to guide the management process in the network elements to adopt the appropriate management behavior and to identify the constraints during the optimization of the utility functions. Some parts of goal policies are translated into abstract policies called behavioral policies, in order to help the autonomic entity to manage its own behavior in accordance with them. The last form of policies aims to describe the behavior the network equipment is to follow in order to reach the global desired state of the network. ANEMA aims at supporting all these mechanisms to achieve autonomic IP network management architecture.

AUTONOMIC NETWORK MANAGEMENT OVERVIEW

The central problem in system and network management lies in the fact that the critical human intervention is time-consuming, expensive, and error-prone. The autonomic computing initiative [3] [6] (launched by IBM) seeks to address this problem by reducing the human role in the system management process. This initiative proposes a self-management approach that can be described as the capabilities of the autonomic systems and the underlying management processes to anticipate requirements and to resolve problems with minimum human assistance. To do so, the human role should be limited; the human should only have to describe the high-level requirements and to delegate control tasks that the system performs in order to achieve these requirements. The associated approach is generally called the autonomic or goal-oriented management approach [7]. This approach starts with the specification of high-level requirements. Then, these requirements are analyzed, refined, composed, and formulated into implemented policies according to the target system-specific technologies [7]. These operations present the most important tasks in the specification and the implementation of the autonomic management approach. They have recently been the focus of several research studies. Hence, many studies were interested in developing methods performing analysis and refinement of high-level goals into low-level policy [8, 9]. Other studies have also addressed the goal decomposition issue [10]. In order to enable multiple constituencies that have different concepts and terminologies associated with the multiple policy description views and to co-define policies, a policy continuum concept has been defined [11, 12]. This concept consists of a stratified set of policy languages, each employing terminology and syntax appropriate for a single constituency [11].

Considering that the policy-based management approach is a good basis on which autonomous decision-making processes for network management can be developed [13], the next issue to consider is obviously how policies can be used to guide the decision-making process in autonomic network management. Walsh and Kephart [14, 15] have addressed this issue, and in their contribution, they have identified several policy concepts in the autonomic system:

- An action policy is expressed in the form: IF (*Conditions*) THEN (*Actions*), where *Conditions* specify a specific state or set of possible states that all satisfy the given *Conditions*, and *Actions* depict the actions that the system should take whenever it is in a state specified by *Conditions*.
- A goal policy is a specification of either a single desired state or criteria that characterize an entire set of desired states. Hence, the goal policies provide only a binary state classification: "desirable" and "undesirable."
- A utility function policy is an extension of goal policies: rather than performing a binary classification into desirable versus undesirable states, the utility function attributes a real-valued scalar desirability to each state. The utility function concept was originally used in microeconomics [16].

Each of these policy concepts is used separately in existing network management solutions (e.g. [15, 17]). We believe that only the integration of all these policy concepts in the management process can help a network to exhibit autonomic behavior. In addition, we propose to extend this set with the "behavioral" policies. Those policies not captured by other policies aim to help the autonomic entity to manage its own behavior depending on its context information. In the following section, we describe in more detail our autonomic network management architecture (ANEMA) and particularly how it instruments all these policy concepts.

ANEMA: ARCHITECTURE AND CONCEPTS

ANEMA defines an autonomic network management architecture that instruments a set of policy concepts to achieve autonomic behavior in network equipments while ensuring that the network behavior as a whole is achieving the high-level requirements from human administrators as well as the users (designed by the human operators in Figure 4.1). As this figure shows, the ANEMA is organized into two layers: an objectives definition layer and an objective achievement layer.

- *Objectives definition layer*: The main component in this layer is the ODP (*Objectives Definition Point*). It allows the administrators to introduce their high requirements and the experts to introduce high-level management

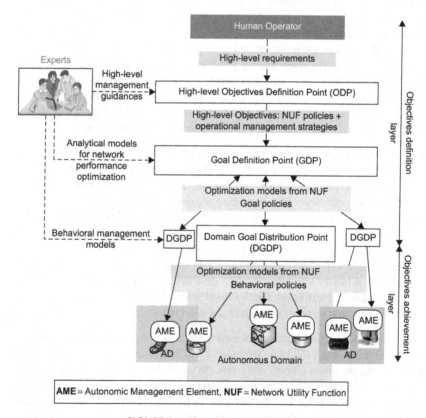

FIGURE 4.1   Global View of ANEMA.

guidelines. In the ODP, the high-level requirements are transformed into *Network Utility Functions* (NUF), whereas the management guidelines are transformed into abstract management strategies. The NUFs represent the policy rules that describe the network performance criteria from the human viewpoint. They are used to express the network functionalities in terms of optimization functions; whereas,the management strategies allow description of the high-level guidelines needed to map the NUF to a specific management architecture that can be implemented within the target network infrastructure.

The ODP forwards the formulated NUF and strategies to the GDP (*Goal Definition Point*), which analyzes them and identifies the related goal policies. In fact, the GDP analyzes the NUF and selects the appropriate global models related to optimization of the NUF metrics from the quality metrics optimization models introduced by the experts. The GDP also selects the adequate management strategies needed to achieve the NUF. The selected strategies represent the goal policies, which are defined as an aggregation of management strategies needed to conduct specific network architecture to achieve one or more quality metrics related to the target NUF expressed at this stage in terms of analytical optimization models. To facilitate the goal policies distribution, the network infrastructure is divided into multiple and smaller domains. Each domain is supervised by a coordinator (called DGDP: Domain Goal Distribution Point). The GDP forwards the goal specifications and global NUF optimization models to all DGDPs. Each DGDP analyzes these specifications and identifies the behavioral policies according to goal policies by using several abstract behavioral management models introduced by the experts. The last form of policies is expressed in terms of behavioral management rules recognized by the target AMEs (Autonomic Management Entities), also called GAPs (*Goal Achievement Points*), in various management situations. The global NUF optimization models are transformed into specific models by applying the constraints deduced from the goals. Afterward, the DGDPs send the behavioral policies and the specific NUF optimization models to all AMEs that are under its control.

- *Objectives achievement layer*: this layer contains a set of GAPs. Each GAP is an entity that behaves in an autonomic manner while trying to achieve the target high-level requirements by considering the goal policies and the NUF optimization models. In fact, according to these informational elements, the GAP can take its own decisions to achieve the target requirements by means of its elementary monitoring, analyzing, planning, and executing (MAPE) capabilities. The GAP is also able to interact with its environment and communicate with other GAPs. In real networks, a GAP can be a router, switch, gateway, software, multimedia device, and so on.

It is noted that experts only need to introduce the knowledge in the knowledge base of ANEMA components but are not part of the management process.

For this reason, the link between the experts and the management components is schematized by dotted lines in Figure 4.1.

NUF and Management Strategies Specification

A NUF describes the utility of the network from the human perspective. Hence, the NUF captures the high-level human requirements expressed in terms of optimization functions (e.g., maximize the network resources utility'). Each NUF is described with a set of quality metrics. The evaluation of the NUF is generally based on an analytical utility function associated with these metrics.

This evaluation allows specifying a space of feasible states. The optimal state in this space is the one that offers maximum network utility. This state is designed, in Figure 4.2, by an optimal feasible state. The management strategies described in the goals are used to determine the valuable and the more desirable state in this space.

By considering only the NUFs related to the self-configuration and the self-optimization capabilities, the guidance of a network to achieve these NUFs requires the specification of two types of management strategies: configuration strategy and optimization strategy. The aims of configuration strategies is to guide the network and its AMEs to adopt the appropriate configurations in a particular context in order to help them achieve high-level requirements. Here, the context is defined as any and all information that can be used to characterize the situation of the entity in its environment [18]. The configuration strategies are transformed into behavioral management policies specified by an ontology language. These policies provide precise definitions of required actions that the AME should follow to reach the desired state. Optimization strategies define

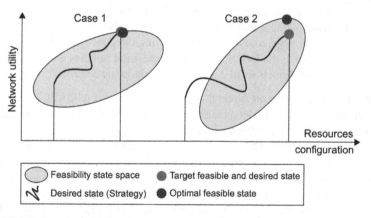

FIGURE 4.2   State Space Description. This figure illustrates two cases of specification of the feasible target and desired state. In case 1, feasible and desired states are the same in case 1 but different in case 2.

the guidelines that the network follows to optimize its utilities. These strategies are translated into analytical optimization constraints used to identify the valuable state among the feasible state space specified by the analytical target NUF resolution that illustrates Figure 4.2. Although they represent different roles in the management process, the configuration strategies and the optimization strategies are dependent, as described in the following subsection.

Goal Policies Specification

The same NUF could correspond to different combinations of management strategy as depicted in Figure 4.3. This figure shows that the NUF has several quality metrics. An elementary utility function (EUF) is associated with each metric. The achievement of each EUF is described according to a set of configuration strategies. Choosing a specific configuration strategy to achieve each EUF can identify the number of optimization strategies able to be applied to optimize the quality metric related to the EUF. Therefore, the goal policies are reformulated by combining the configuration and optimization strategies that can be applied to achieve the NUF, a set of EUF (the case of G4 in Figure 4.3), or a particular EUF (the case of G1, G2, and G3 in the figure). The network can adopt different goals to achieve the same NUF. In order to ensure the convergence of the network to the desired state defined by the goal, a cooperation process between the AMEs is desirable, but it is not addressed in this contribution.

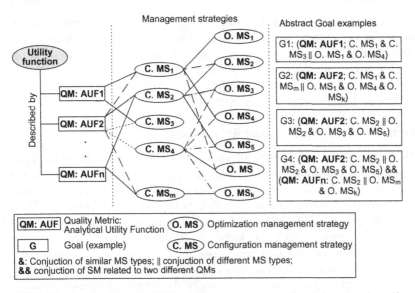

FIGURE 4.3 Goal Specification. This figure illustrates some examples of abstract goal specification related to an NUF described by quality metrics.

Hence, at some stage it is also necessary to address conflicting goals in an AME when this AME is associated with different domains or authorities.

Behavioral Policies Specification

Behavioral policies are organized by policy action groups whereby each group describes a behavior (therefore they are called BPs—Behavioral Policies). Each behavior is related to a particular AME profile and a particular management context. We have structured the behavioral policies in four parts [19]:

- *Context part* specifies the management context description.
- *Configuration part* describes the set of rules needed to configure and organize the infrastructure regarding context properties.
- *Monitoring part* describes the set of rules needed to implement the monitoring processes in order to control the AME environment.
- *Event-Condition-Action (ECA) part* specifies the abstract ECA rules. Each rule includes three elements: (1) *Event*: An event specifies the signal that triggers the invocation of the rule. (2) *Conditions*: The conditions represent a set of tests, if positively verified, that cause the execution of the actions. (3) *Actions*: The actions consist of a set of actions to react to the event effects. This element is also used also by the configuration and the monitoring parts. In these cases, the actions are applied to effect a configuration or to launch the monitoring process. The actions can be of different types such as communication action, behavior action, configuration action and monitoring action.

The behavioral policies do not specify any parameters or implementation instructions in order to allow the AMEs to adapt their behaviors according to their internal states. The limitation of these behavioral policies descriptions resides in their limited capacity to define autonomic behavior in AME. One approach that can be followed in the future is use of reasoning capabilities such as the case-based reasoning method [20].

Policies Transformation and Relationship

The transformations related to all views of our proposed architecture represent a policy continuum, as introduced in [11, 12, 21]. Figure 4.4 illustrates the ANEMA policy continuum that is composed of five views. This continuum starts with the Business view, which defines policy in business terms only (SLAs and high-level requirements). The System view translates the business policy into system concepts (NUF and high-level management strategies) neutral to any technology. The Network view maps these neutral system specifications into models related to the technology specifications. The AME view defines the specific types of devices used to describe the behavioral policies, while the Instance view takes into account the specific executable policy and configuration commands related to description of the behavioral policies and the device parameters. Table 4.1 summarizes the main concepts used in the ANEMA policy continuum:

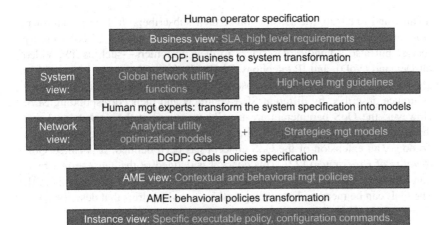

FIGURE 4.4   ANEMA Policy Continuum.

TABLE 4.1 ANMA concepts Description

Objective	An objective is a management rule expressed in a high abstract level. In our architecture, the objectives are composed of utility function policies and management strategies.
Network Utility Function (NUF)	A NUF represents the utility of the network from the human operator's viewpoint. This function is described in analytical optimization form. Its evaluation allows specifying the feasible states space.
Management strategy	A strategy is a description of a high-level guideline needed to conduct an entity (network or equipment) to achieve one or more quality metrics characteristics.
Goal	A goal is an aggregation of management strategies needed to conduct specific network architecture to achieve one or more quality metrics related to the target NUFs. A goal also aids in determining the desired state in the feasible states space.
Behavioral Policy (BP)	A BP is a set of abstract policies describing the AME behavior according to a specific context description.
Context	A context is any and all information that can be used to characterize the entity environment situation [18].

AUTONOMIC QOS/QOE MANAGEMENT IN MULTISERVICE IP NETWORKS

To illustrate how ANEMA can be used in a real network, we have selected the case study where operators (designated by NP: Network Provider) adopt the ANEMA solution, willing to control their IP network in a more efficient manner and to ensure the satisfaction of their subscribers. In fact, actual operators are using different technologies to provide not only an IP connectivity service but also a bundle of services to their customers—such as TV, Video on-Demand (VoD), and IP telephony (VoIP). In this context, service providers are willing to provide different subscriptions with different quality of service assurance described by different SLA contracts and their corresponding SLS. However, the QoS parameters are usually measured at the packet level and therefore do you capture properly the quality of experience of the subscribers. To do so, an extension of the QoS concept has been defined; it is called QoE (*Quality of Experience*) [22]. This concept is defined as the cause–effect relationship among quality characteristics that might affect user perceptions [23]. The QoE can be measured according to the network metrics that describe these characteristics [23]:

$$<Begin>$$
$$QoE_K = f(metric_1; \ldots, metric_{nk})/k = 1 \ldots K$$

with K: the number of considered quality characteristics and nk: the number of metrics required to describe the kth quality characteristic.

To meet all users' requirements and to guarantee the requested QoS, the operator has to deploy a very complex monitoring and control infrastructure. To reduce this complexity, ANEMA can be used to provide an autonomic management of the QoS and the QoE in the network. For the sake of simplicity, we have only considered the bandwidth (designed by Bw in the rest of this chapter) as the main parameter to optimize. In the following subsections, we describe the main steps to deploy the ANEMA solution.

Step 1: Identification of High-Level Requirements and NFU

The multiservice IP network architecture introduces mainly two actors: customers and network administrators. The NPs have a global view of the network optimization (i.e., ensuring the highest global QoS while optimizing the network resources utilization, in order to *ensure the best return on investment*. From the customer's perspective, their objective is to *have a good experience with their preferred services while reducing their bill.* These two objectives represent the high-level requirements to be achieved by the autonomic network based on ANEMA. Analytically, these optimizations are translated into the following utility functions: (1) maximize the NP revenue and (2) optimize the customer's preferred services quality. From the network perspective, the customer preference is translated to a QoE optimization considering the customer' preferences. These two objectives need to be optimized conjointly' therefore, the NUF can be expressed as follows:

max(QoE) & max(NP revenue) considering the customers' preferences

To resolve this NUF, we formulate an analytical model considering only the Bw allocation metric. The most important formulas and metrics used in this model are described in the following subsection.

Step 2: NUF Analytical Description

The analytical model used to describe and to resolve the NUF employs a set of predefined Traffic Utility Functions (TUF) related to the Bw allocation metric. This model consists of calculating two service utility metrics: the Expected Utility (EU) and the Perceived Utility (PU). These metrics are used thereafter in estimating the QoE and NP expected revenue [24].

Predefined TUF Related to the Bw Allocation

As described in [25, 26], the traffic crossing the network can be differentiated between three traffic patterns: hard real-time (also called CBR: Constant Bit Rate) traffic, adaptive real-time (called also VBR: Variable Bit Rate) traffic, and elastic (called also UBR: Unspecified Bit Rate) traffic. A single UF ($U(bw)$) is defined for each one of these traffic patterns.

These TUFs are used to calculate the average connection utility (V) as follows:

$$V_i = \frac{1}{T_{dur}} \int_0^{T_{dur}} U_i[bw_i(t)]\,\mathrm{d}t \text{ where } T_{dur} \text{ is the duration of the session.}$$

This utility represents an approximation of the network session performance as perceived by the customer during a certain period of time [26]. For this reason, it was chosen to estimate the QoE and then to calculate the NP revenue. However, in our case study, the customers' preferences must be taken into account in valuating these metrics. For that purpose, we have introduced a new parameter (called SPref[1]) that represents the importance assigned by the customer to each service according to his preferences.

Expected Utility Estimation

The service's EU is calculated by resolving the following optimization problem: maximizing the sum of customers' activated service utilities by considering their SPref parameters during time and according to the available users' Bw. Analytically, this problem can be formulated as follows:

$$\max_{bw \geq 0} \sum_{i=1}^{k} SPref(T_i)^* U_{T_i}(bw_i) \text{ subject to } \sum_{i=1}^{k} bw_i \leq B_u \qquad (1)$$

where k: = the number of activated services from the user, $SPref(T_i)$: = the ith service importance degree, bw_i = the estimated Bw to the ith traffic, and finally, Bu = the customer's access link capacity (generally specified in the *Service Level Agreement* (SLA). The *EU* is equal to: $EU_{Ti} = U T_i(bw_i)$ where bwi is the ith component of the Equation (1) solution vector.

---

[1] SPref is a scalar value $\in [1\ldots QL]$ with QL the number of QoS levels offered by the network.

Perceived Utility Measurement

In order to provide a high QoS to the users as well as to maximize their benefits, the NP must maximize the services utility by considering their importance for the users. To fulfill this goal, the service's PU is defined as the utility offered by the network to the service's traffic flow. This utility is equal to: $\min_{j=1...L} U_{T_k}(bw_j)$ where L = the number of links traversed by the traffic k, bwj = allocated Bw to the traffic k in the link j. To calculate this utility, the Bw repartition management process uses the following formula in order to determine the best Bw repartition of the link capacity (C_{link}) between the N traffic flows present in the link:

$$\max \sum_{i=1}^{N} SPref(T_i)^* U_{T_i}(bw_i) \text{ where } \sum_{i=1}^{N} bw_i \leq C_{link} \tag{2}$$

NUF Evaluation Metrics: QoE and NP Revenue

The *QoE* related to the service i (described by its traffic type Ti) is calculated using the EU and PU relative to this service, as follows:

$$QoE(T_i) = \frac{1}{T_{dur}} \int_{1}^{T_{dur}} a_i(t)\, dt \text{ where } a_i(t) \text{ is described as follows} \tag{3}$$

$$a_i(t) = \begin{cases} 1, & \text{if } PU_{Ti}(t) \geq EU_{Ti}(t) \\ 1 - SPref_{Ti}(t)^*(EU_{Ti}(t) - PU_{Ti}(t)), & \text{if } PU_{Ti}(t) \leq EU_{Ti}(t) \end{cases}$$

The NP revenue that will be gained from the service exploitation is equal to:

$$C_i = QoE^* p_i^* T_{dur} / Time_{unite_{pi}} \quad \text{where } p_i: \text{ the unit } i\text{th service price.} \tag{4}$$

Step 3: Management Strategies Specification

A set of management strategies can be defined to lead the network achieving the NUF described in Step 1. By considering only the Bw allocation and repartition metrics, these strategies are related to the QoS architecture, the Bw repartition per class of services, and so on. Table 4.2 presents some examples of Bw management (allocation and repartition) strategies:

Step 4: Goals Specification

In our case study, the goals describe the selected management strategies for managing the network and optimizing its performance. By considering the examples of management strategies described in Step 3, a set of goals can be adopted by each GDP. Table 4.3 presents two examples of goals among this set. The configuration strategies in these goals are transformed into BP models that we describe in the following step. The constraints related to the optimization strategies are applied to the NUF resolution model (described in

TABLE 4.2 Examples of Bw Management Strategies

Type	Description	Example
Configuration	QoS architecture strategy	Differentiated Services architecture; Management Context: Number of traffics that need different treatments.
Optimization	Bw repartition strategy	No Limited Bw repartition per class of services and 100% Bw allocation of the first level requests

TABLE 4.3 Examples of Goal Description (in these examples we suppose that the traffics are differentiated between three classes of services. QL1 = Quality level 1/Spref = 3)

Goal 1	Goal 2
Configuration strategy: DiffServ; Management context: Number of traffics that need different treatments	Configuration strategy: DiffServ; Management context: Number of traffics that need different treatments
Goal Optimization: Limited Bw-Repartition (BR) per classes of services: $Bw(Cli) < \alpha_i * C / i = 1 \ldots 3$, $C =$ link capacity Initial BR: $Bw(Cli) = Bwlni(Cli)/I = 1 \ldots 3$ where $Cli = i$th class of services, α_i: Cli % Bw portion	Goal Optimization: No Limited Bw-Repartition (BR) per classes of services and assuring 100% Bw allocation to services that have the highest importance for their users: $0 < Bw(Cli) < C / i = 1 \ldots 3, C =$ link capacity Bw(Cli1) $$= \$ \sum_{k=1}^{N} (bw_{ik_{QL1}})/N$$ services having the SPref = 3 in Cli. Initial BR: $Bw(Cli) = Bwlni(Cli)/i = 1 \ldots 3$

Step 2). For example, by considering the optimization strategy defined in Goal 1, Equation (2) becomes:

$$\max \sum_{i=1}^{3} \sum_{j=1}^{n_i} SPref(T_{ij}) * U_{T_{ij}}(bw_{ij})$$

$$\text{where} \quad \sum_{j=1}^{n_i} bw_j \leq \alpha_i * BW \quad \text{and} \quad \sum_{i=1}^{3} \alpha_i * BW = C \tag{5}$$

Step 5: Behavioral Policies Specification

The configuration strategy described in the last goal examples aims to guide the network to efficiently manage the heterogeneous traffics using the Differentiated Services (DiffServ) [27]. In DiffServ model, the traffics are classified in different classes in order to provide various QoS for each class of services. Since the traffics are classified according to their types, the number of classes needed to differentiate all traffics flooding in the network at any given time depends on the number of types of these traffics. For this reason, in order to optimize the number of classes to configure, the adopted context is described using the number of traffics needing different treatments (class of services). By considering this context and by limiting the number of classes to 3, we can describe three different situations: (1) all traffics need the same treatment (one class of services is needed to classify all traffics), (2) all traffics need two different treatments (two classes of services are needed to classify all traffics), and (3) all traffics need three different treatments (three classes of services are needed to classify all traffics). For each situation, a specific behavior is described (BE: Best-Effort, PQ: Priority-Queuing and DiffServ: Differentiated Services, respectively, to the last situations) (Figure 4.5). Each behavior is represented as a state and a specific event (Event-oldstate-targetstate-transition) as a transition between two states. Therefore, the BP model has a form of state-transition diagram as illustrated in Figure 4.5.

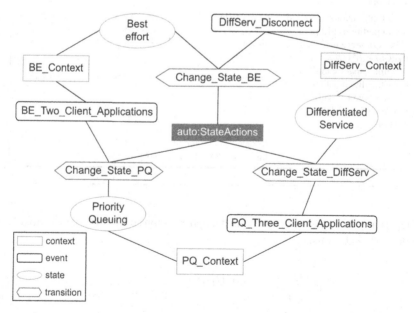

FIGURE 4.5 Description of the BP according to the management context (State-transition Diagram).

Technical Specification of the GAP: Autonomic Router

The router is part of the equipment in which the management goal can be enforced in the network; its role is to achieve the goal policies locally. These policies, expressed in terms of analytical and abstract models, define abstract conditions and configuration actions. The autonomic router translates the abstract configuration parameters into local parameters depending on its own capacities. In order to allow the autonomic router to achieve this role, we have proposed evolving its internal architecture. The proposed architecture is aligned with the one presented in [2] and is composed of a number of functional modules that enable the router to exhibit the expected behavior. These modules are as follows (as depicted in Figure 4.6):

1. *Control Module* that introduces two entities: (a) the sensors that provide mechanisms to monitor the environment of the router, and (b) the effectors that allow configuring the router managed resources. In our work, the control module affects mainly the configuration of the underlying traffic engineering mechanisms of the IP router using our own API (*Application Programming Interface*) based on the Linux Traffic Control (TC) tool [28]; (2) *MM: Monitoring Module* provides different mechanisms to collect, aggregate, filter, and manage information collected by sensors. (3) *AM: Analyze Module* allows diagnosing the monitoring results and detects any disruptions in the network or router resources. This information is then transformed into events; (4) *PM: Planning Module* defines the set of elementary actions to perform according to these events; (5) *EM: Execution Module* provides the mechanisms that control the execution of the specified set of actions, as defined in the planning module. In addition to these modules, this architecture introduces a Policy Manager Module (PMM) that transforms the abstract BP action into a plan of executable actions deduced from the specific router information models saved in its local database.

Algorithm 4.1 specify the main behavior of the management router.

The objectives of this approach are (1) to maximize the client's satisfaction and (2) to guarantee a good ROI for the operator. However, the proposed approach is purely analytical, so it only permits analytically calculating the perceived quality and the services exploitation costs. In order to effectively implement these utility functions that will enable a network administrator to delegate the decision-making process to a set of network elements we need to design and implement these metrics using an open, yet rich information model. The following section describes the extensions we have made to the DMTF Common Information Model in order to implement our analytical approach.

QOE INFORMATION MODEL DESIGN

The DMTF issued a set of specifications that can be used to implement a standardized framework for the management of systems, users, metrics, networks,

---

Algorithm 4.1 Behavioral Management in Autonomic Router

Input: specific Behavior

1: Behavior-specific Behavior **do**
2: **while** Behavior-specific Behavior **do**
3:     **for** all Behavior.Config-Action and all Behavior.Monitor-Action **do**
4:         Action Behavior.Config-Action or Behavior.Monitor-Action
5:         AM analyzes the Action.type and specifies the Action.parameter-Vector.
6:         PM, in cooperation with the PMM, identifies the Action.elementary-actions according to the Action.parameter-Vector.
7:         EM transforms Action.elementary-actions in a plan of executable actions (using TC API) to be executed by the effectors.
8:     **end for**
9:     Event ''
10:     **while** !Event **do**
11:         Sensors monitor the monitoring processes associated to the performed Monitor-Action.
12:         MM collect the results of monitoring processes provided by the sensors.
        13: AM
13:         AM diagnosis these results and identifies the eventual produced Event.
14:     **end while**
15:     EM commands the effectors to stop the performed monitoring processes.
16:     AM analyzes the ECA. Conditions associated to Event and compares them to the router parameters.
17:     **if** ECA.Conditions are verified **then**
18:         AM analyzes the ECA.Actions.
19:         **for** all ECA.Actions **do**
20:             Action ECA.Actions
21:             **if** Action.type = ChangeBehavior **then**
22:                 Behavior = newBehavior
23:             **else**
24:                 AM analyzes the Action.type and specifies the Action.parameter-Vector.
25:                 PM, in cooperation with the PMM, identifies the Action.elementary-actions according to the Action.parameter-Vector.
26:                 EM transforms Action.elementary-actions in a plan of executable actions to be executed by the effectors.
27:             **end if**
28:         **end for**
29:     **end if**
30: **end while**

---

and the like. The DMTF Common Information Model (CIM) is the management information model used to standardize a common conceptual framework for describing the managed environment. Once the CIM providers (the platform-specific implementation of the CIM model, or a part of it) are provided on a given environment, the system administrators can use the Web-Based Enterprise

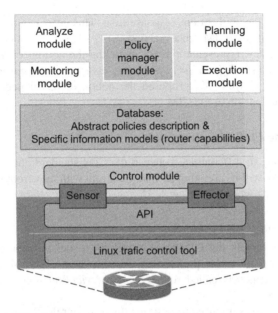

FIGURE 4.6 Proposed Architecture for ANEMA Autonomic Router.

Management (WBEM) technologies to effectively manage the networked element via the standard CIM interfaces. The objective is to implement a common management interface for all types of resources without having to handle various information model and proprietary protocols. It all works like a Client-Server model where an administrator can use a single management application (client) to control its network resources as long as the providers are available via the standard CIM Object Manager (server).

In order to represent the service's utility function and the user's preferences information, we have chosen to extend some part of the CIM model, keeping in mind the implementation of these extensions (QoE CIM providers) whenever our approach is validated.

Despite the large number of specifications available in the CIM, the concepts we will be using for QoE quantification cannot be implemented with the current release of the DMTF information model. Thus, we have extended the CIM definitions to include our original contribution for (1) a specific user preference declaration, (2) the notion of utility function to represent the traffic (application), (3) as well as the performance metrics calculation based on a utility function.

The CIM is an open information model, and the DMTF provided many extension mechanisms to facilitate its enhancement by vendors' specific concepts (metrics, software/hardware characteristics). Many contributions tried to enforce the new metrics definitions in the CIM. In [29], the authors extended the CIM metrics model to allow the performance calculation of Delay-based Congestion Avoidance algorithms for different versions of TCP. This formalism was reused by [30] to describe another CIM extension that permits the measurement of the applications response time. This latter extension was officially included in the CIM since version 2.17. Also, the work described in [31] introduces new representations for modeling the notions of metering and accounting (*provisioning, capacity planning, auditing, and charging the customers*). Finally, Doyen [32] proposed the notion of distributed hash tables (DHTs) as an extension of the CIM.

Modeling the QoE Measurement

In the following, to differentiate between the CIM standard concepts and our contributions we will be prefixing each new concept with the acronym ANEMA, which stands for *Autonomic NEtwork Management Architecture*.

Modeling User Preferences for QoE

This extension targets the CIM user model by associating the concepts of user, service and its usage preferences. As shown in Figure 4.7, we introduce two new concepts called *ANEMA_UserServicePreference* and *ANEMA ServicePreference*.

These two classes are used to describe the users' preferences for the usage of a particular service. The concept *ANEMA_ServicePreference* defines the main SPref managing function "*Identify*" to specify an SPref, "*Modify*" to update it, and "*Delete*" to remove an existing one.

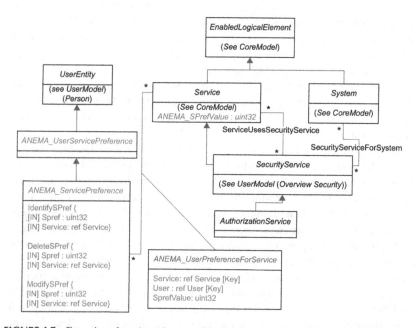

FIGURE 4.7   Extension of user's preferences (SPref) in the CIM User and CIM Service Domains.

In order to establish a relationship between the user and the service concepts, we propose an association class called *ANEMA_UserPreferenceForService*. This association class establishes the SPref value (*ANEMA SPref Value*) attributed by a user to a service. This attribute will be used to specify the service class as part of the Differentiated Services Code Point (DSCP) range (precedence bytes).

Modeling Quantification Metrics for the QoE

In order to represent the methodology behind the usage of utility functions for measuring the QoE, we started by defining units of work that characterize the generic operations for the QoE determination. Then, we assigned a set of metrics to these units of work. All these necessary extensions have been added to the CIM information model.

● Definition of units of work during the utility-based calculation of the QoE

The QoE calculation relies on a set of elementary processes described in the following:

(1) The calculation of the user-estimated utility: This process consists in calculating the estimated utility for a service or a set of services with regard to the available bandwidth (for the user) while taking into account the SPrefs values of each active service.
(2) The calculation of the network-related utility: This process allows the calculation of the service, or a set of services, utility with regard to the network availability.
(3) The calculation of the service-related QoE: This process permits quantifying the service, or a set of services, QoE, while taking into account the user-estimated utility as well as the real-time network-related utility.
(4) The calculation of the real-time bit-rate for an application: This process is responsible for calculating the real-time bit rate of each application.

These four processes are essential for measuring the service performance with regard to the user expectations (via the QoE_{rate} metric). The identification of these processes helped define the four main work units: **ANEMA_EUUoWDefinition, ANEMA_PUUoWDefinition, ANEMA_QoEUoWDefinition and ANEMA_BitrateUoWDefinition.** These processes are shown in Figure 4.8. The following describes how we have modeled the execution of these processes.

The QoE Quantification Processes

The modeling of the QoE_{rate} quantification consists in abstracting its execution steps in order to facilitate the identification of the associated metrics as well as its measurement parameters. In order to integrate the data provided by these metrics in our management information model, we have formalized them in terms of units of work and metrics in the CIM.

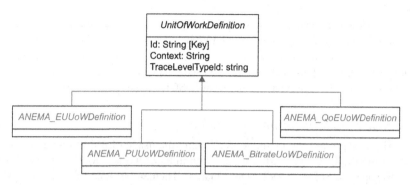

FIGURE 4.8   Unit of Work Definitions for QoE Quantification.

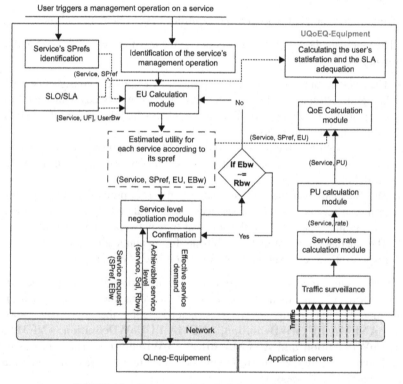

FIGURE 4.9   Abstraction of the QoE Quantification Process.

As shown in Figure 4.9, the QoE quantification process starts by an interaction between the client (user) and its associated QoE measurement equipment (noted as ***UQoEQ-Equipment***). This first interaction is for enabling a management operation (activation, deactivation, priority/importance update) on a service or a set of services. The ***UQoEQ-Equipment*** is responsible for identifying the user's request and the service importance. The service importance level can be predefined and often remains standard with regard to the users' initial contractual needs (Conferencing ≪ Video watching ≫ Web navigation).

In order to calculate the best bandwidth repartition allocated by the operator, the **UQoEQ-Equipment** queries its local SLA/SLO database to choose the right utility function for each activated service (UF) as well as the allowed bandwidth (**UserBw**). Once the right configuration is identified, the **UQoEQ-Equipment** calls the module responsible for calculating the estimated utility for each activate, or to be activated, service. This module returns the estimated utility for each service as well as the allocated bandwidth along with its importance level—which is important if the system triggers an automatic adjustment of the service importance levels. Once the estimated utility (EU) is calculated, the quality-level module initiates the negotiation, with the equipment managing the quality level of each service (**QLneg-Equipment**) designated at the other side of the network. This negotiation starts by querying the service with its predefined level of importance (**SPref**) and the estimated bandwidth (**Ebw**). Considering these initial service requirements, the **QLneg-Equipment** replies by indicating to the **UQoEQ-Equipment** the possible service level (Service Quality Level—SQL) and the required bandwidth (Rbw)). Then, the **UQoEQ-Equipment** verifies the concordance between the **Ebw** and the **Rbw**. In case of concordance, the **UQoEQ-Equipment** confirms the **QLneg-Equipment** request and stores the EU for this service. In case of incompatibility, the **UQoEQ-Equipment** readjusts the SPrefs to meet the provider's service offer. In some cases, this readjustment will necessitate the client's approval when the negotiation module cannot proceed without fulfilling the initial users' preferences—in case of inappropriate SPrefs levels or too many activated services.

The second part of the process consists in controlling the incoming traffic in order to recalculate the bit rate for each service considering its effective value. Periodically, the PU calculation module measures the utility offered by the network for each service. This utility is used by the QoE measurement module. At each closed session, this module sends a notification to the client's satisfaction calculation module (the corresponding QoE for this session).

Metric Definition and Modeling

As shown in Figure 4.9, the QoE measurement process consists in three elementary processes. The first calculates the estimated utility, the second permits calculation of the perceived utility by the user, and the third process computes the results of the previous processes to determine the QoE level—thus, the user satisfaction level.

In order to model our approach for quantifying the QoE, we propose characterizing the parameters of each elementary process by defining the following metrics:

- **Utility function:** This metric permits description of the utility function for each service. The utility function depends on a set of parameters, such as minimal banwidth (Bw_{min}) and the partition parameter. To implement this function, it is necessary to identify an additional metric that is related to traffic type(*trafficType*))

- **UserBw:** This metric permits specification of the bandwidth that is really allocated to the client by its operator. This metric is an SLA parameter.
- **Estimated Utility (EU):** This metric permits description of the user's estimated utility measure for one of its services. It is a scalar value comprised between 0 and 1.
- **Perceived Utility (PU):** This metric is used to describe the user's perceived utility measure for one of its services. It is a scalar value comprised between 0 and 1.
- **Quality of Experience (QoE):** This metric permits description of the service's effective utility measure. It is a scalar value comprised between 0 and 1.
  - **Bit Rate:** This metric describes the traffic's bit rate at the user's connection point (the traffic's bit-rate reception).
  - **Traffic Quality:** This metric permits measurement of the traffic's quality from the perceived utility calculated value.

Each metric definition was modeled as a subclass of the *CIM MetricDefinition,* which is a subclass of *CIM_BaseMetricDefinition.* As shown in Figure 4.10, these subclasses are called **ANEMA_UFDefinition, ANEMA_EUDefinition, ANEMA_PUDefinition, ANEMA_BitRateDefinition, ANEMA_QoEDefinition** and **ANEMA TrafficQualityDefinition**.

Following the CIM pattern, the instantiation of each metric is realized via the **ANEMA_MetricValue**, which is a subclass of *CIM_BaseMetricValue.* Figure 4.11 illustrates instantiation for the utility function **ANEMA_UFDefinition.**

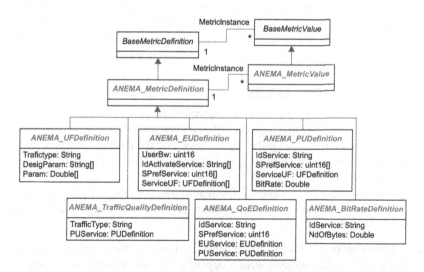

FIGURE 4.10   Metrics Definition for the QoE Quantification.

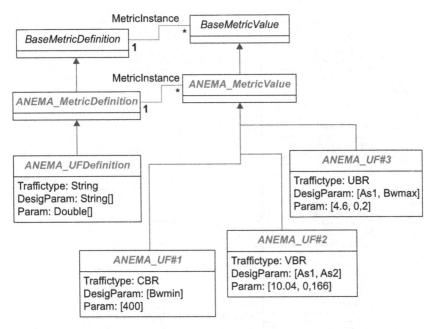

FIGURE 4.11   Sample Instances of the **ANEMA_UFDefinition** Class.

Three types of traffic are described: Variable Bit Rate (VBR), Constant Bit Rate (CBR), and Unspecified Bit Rate (UBR).

Modeling Work Units for Measuring the Bit Rate-Related QoE:

The model for quantifying the QoE is illustrated in Figure 4.12. This regroups all the units of work and the metrics previously defined in this section.

The **ANEMA_QoEUoWDefinition** class is the QoE measurement task executed by the *UQoEQ-Equipment*. These classes are associated via the *LogicalElementUnitOfWorkDef* association, which permits one to assign a unit of work to the equipment that executes it.

As previously stated, the QoE measurement task encompasses two main actions: measuring the estimated utility and measuring the perceived one. Each action is modeled as a unit of work: **ANEMA_EUUoWDefinition** and **ANEMA_PUUoWDefinition** classes. Instances of the *CIM_SubUoWDef* association reflect their imbrications to the **ANEMA_QoEUoWDefinition** class.

The corresponding class is related to the **ANEMA_PUUoWDefinition** class via the association *CIM SubUoWDef*. It is also linked to the **ANEMA_UQoEQ-EquipmentInterface** class via the *LogicalElementUnitOf- WorkDef* association.

All the units of work are related to their corresponding metric via the *CIM_UoWMetricDefinition* association, whereas the dependence of one metric on another is described by the ANEMA UseMetric association. For example,

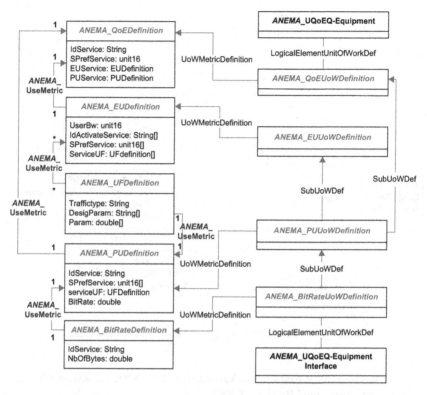

FIGURE 4.12 Modeling the Work Units for Measuring the Bit Rate-Related QoE The unit of work for measuring the perceived utility encapsulate the real bit-rate calculation for each traffic from the **UQoEQ-Equipment** interface. This activity is also described in terms of unit of work (**ANEMA_BitrateUoWDefinition**).

as previously stated, the **ANEMA_QoEDefnition** metric depends on the **ANEMA_EUDefnition** and **ANEMA_PUDefnition** metrics.

EXPERIMENTATIONS AND SIMULATIONS RESULTS

In this section, we describe the results of the simulations and experimentations of an ANEMA in the context of multiservice IP networks.

Simulations and Analytical Results

To evaluate the advantages of the proposed analytical model resolution of the NFU in case of a multiservice IP network (as described in the previous section), we have developed a simulator (with MATLAB [34]) and selected six traffic types for the simulation scenarios. Three QoS level (noted L1/L2/L3) are defined for each traffic type. Table 4.4 describes the main characteristics of these traffic types (the characteristics are similar to the ones used in [35]).

TABLE 4.4 Characteristics of Traffics Used in Our Simulations (where TQ = traffic Quantity, AR = Average rate, ACD = Average Connection Duration, QL = Quality Level, Li = ith quality level)

T_i	TQ or AR	ACD	AR	ACD:L1,L2,L3	AR:L1,L2,L3	$U(b(\text{Mbps}))$
T1	30Kbps (VoIP)	180	30	180	30	$\begin{cases} 1, & \text{if } b \geq 0.03 \\ 0, & \text{if } b < 0.03 \end{cases}$
T2	256Kbps (VisioConf)	300	256	300	256	$\begin{cases} 1, & \text{if } b \geq 0.25 \\ 0, & \text{if } b < 0.25 \end{cases}$
T3	120Mb	600	200	200/400/600	600/300/200	$1 - e^{\frac{-10.04*b^2}{0.166+b}}$
T4	20Kbps (Email)	30	20	10/30/60	60/20/10	$1 - e^{\frac{-4.6*b}{0.02}}$
T5	512Kbs (DoD)	180	512	90/180/300	1024/512/300	$1 - e^{\frac{-4.6*b}{0.5}}$
T6	100Mb (File transfer)	240	420	120/240/480	840/420/210	$1 - e^{\frac{-4.6*b}{5}}$

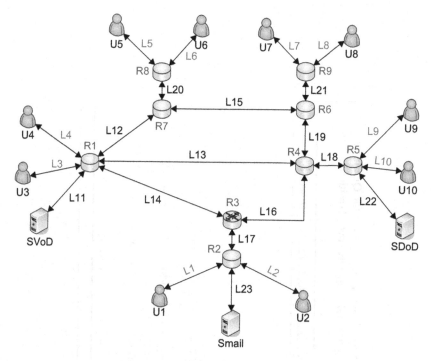

FIGURE 4.13   Simulated Network Topology.

These traffics are divided into three classes in the following way: T1 and T2 class CL1; T3 class CL2; T4, T5, and T6 class CL3. In order to describe the analytical results for the proposed model, several simulation scenarios have been developed and are described in the following.

The scenario consists of a network composed of 23 links (10 1,8Mbps bandwidth links are used to connect 10 users; the 13 core network links that have a individual capacity of 2,7Mbps except L1 and L8 that have respective capacities 3.24Mbps and 4.05Mbps). These capacities are shared among 45 connections (corresponding to the 10 users' unicast service applications). Figure 4.13 illustrates the simulated network topology. The users' preferences (SPref) related to these services are illustrated in Table 4.5. Each row in this table represents a traffic type, while each column represents a user.

Figure 4.14 and Figure 4.15 illustrate the main results obtained from the simulation of the analytical optimization model on the simulated network topology. Figure 4.14 highlights the global utility (i.e., the sum of the individual service utility) as well as the global service cost per user while considering 4 Bw repartition strategies:

This figure (Figure 4.14) shows that the global utility and the services exploitation cost vary according to the adopted management strategy. When comparing the results related to the four considered strategies, it is worthwhile

TABLE 4.5 Users' Preferences (SPref) Matrix

	U1	U2	U3	U4	U5	U6	U7	U8	U9	U10
T1	1	1	1	1	1	1	1	1	1	1
T2	1	1	1	1	1	1	1	1	1	1
T2	2	1	2	3	3	1	2	1	1	2
T4	1	2	3	2	1	2	3	3	2	3
T5	1	2	3	2	1	2	3	3	2	3
T6	3	3	1	1	2	3	1	2	3	1

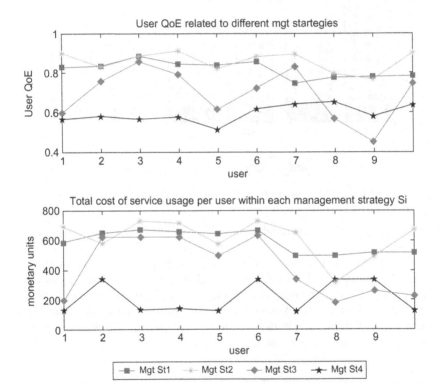

FIGURE 4.14 Global Utility and Service Cost Relative to Four Management Strategies. **Mgt St 1**- No limited Bw repartition tacking into account SPref (standard Mgt St), **Mgt St 2**- No limited Bw repartition without considering SPref, Mgt St 3- No limited Bw repartition considering SPref and assuring the highest quality to services that have SPref = 3 and **Mgt St 4**- Static Bw repartition.

to notice that the first management strategy is the best one. The second figure (Figure 4.15) represents the Bw repartition per the classes of services (CL1, CL2, and CL3) for the 13 core network links by considering the last four management strategies. This figure shows that the Bw repartition is not similar in the

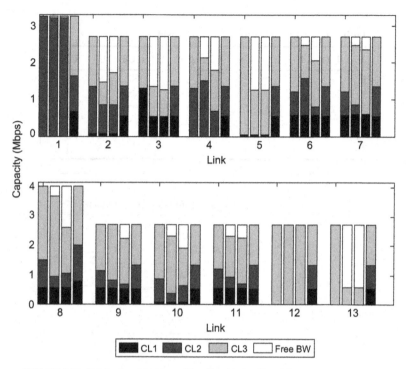

FIGURE 4.15 Links Capacity Repartition Considering Four Management Strategies.

different links, and for a particular link this repartition varies from one strategy to another.

The results shown in this part confirm that the proposed model allows specifying the optimal feasible state (Mgt St 1). However, by considering additional constraints related to the management strategies adopted, the resulting state is still not the optimal one. Although these strategies do not allow optimizing the network performances in the short term, they may involve other favorable effects in the long term (e.g., attract more users).

Testbed and Experimental Tests

In order to experiment with some characteristics of our proposed ANEMA, we have implemented a proof-of-concept testbed representing a multiservice IP network. The testbed topology is composed of four routers, three users' sites, and a service provider. Each customer is allowed to use several services supported by the service provider (services have T1, T2, or T3 traffic types). This testbed topology describes a management domain supervised by a DGDP. To manage this topology, we consider in this testbed (see Figure 4.16) Goal 1 as described in Table 4.3 and the optimization models and BPs described in Steps 2, 3, and 4 of the proposed guidelines in the previous section.

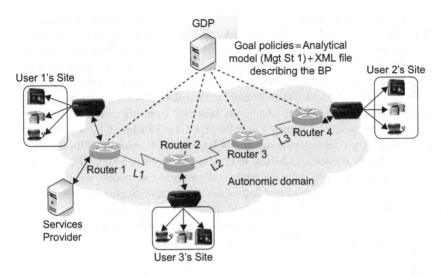

FIGURE 4.16 Description of the Deployed Testbed for Autonomic Network Management with Web, VoIP and VoD services.

In this experimentation, the focus was put on the autonomic router behavior. Therefore only the experimental results related to router behaviors and the core network link capacities repartition are described. As depicted in the BP model described in Figure 4.5, the BP specification defines three behaviors : BE, PQ, and DiffServ.

At the initial stage of our experimental scenario, the routers initialize their behaviors to the default behavior. During this time, the routers monitor their interfaces and their parameters. At time (t1), the Router1 detects the VoIP traffic transiting between user1 and user3. It decides (1) to remain in the BE behavior. This decision is motivated by the fact that the networks is only supporting one type of traffic (T1); (2) to create the class CL1 and to allocate for it 30Kbps from L1 capacity, and (3) to send a message to Router2 to inform him about this event so that they cooperatively recognize this new situation and behave accordingly. By receiving this message, Router2 decides to remain in the BE behavior.

At the same time, Router4 detects the launch of a new application (File Transfer (FT) between user3 and user2). The router4 decides (1) to remain in the BE behavior, (2) to create the class CL3 and to allocate for it all L3 capacity, and (3) to send a message to Router3 to inform it about this new situation. By receiving this message, Router3 decides in its turn (1) to remain in the BE behavior and to create CL3, (2) to allocate for this class 840Kbps from L2 capacity, and (3) to inform Router2 about these decisions.

At time (t2), after detecting a new application (VoD for user2), Router1 decides to change its behavior from BE to PQ and reconfigure L1 as follows: CL1 = 30 Kbps and CL2 = 600 Kbps and inform Router2 about this new situation. To respond to this event (receiving a message from Router1),

Router2 decides to change its behavior from BE to PQ and reconfigure L2 capacity as follows: CL2 = 600 Kbps, CL3 = 840 Kbps and to send a message to Router3. When it receives this message, Router3 changes this behavior (PQ) and recon-figured L3 capacity (CL2 = 600 Kbps, CL3 = 840 Kbps) and trans-mits the message to Router4. At time (t3), user3 decides to launch the FT service with his service provider. When Router2 detects the traffic related to this service in its interface, it decides to change its behavior from PQ to DiffServ, create the CL3, and update L1 capacity as follows: CL1 = 30 Kbps, CL2 = 600 Kbps, CL3 = 840 Kbps. However, some of this Bw portion is equal to 1270 Kbps. Or the L1 capacity is equal to 1 Mbps. To resolve this problem and optimize its per-formance, Router2 performs the analytical optimization models and considers the user's preferences associated with activated services. Supposing that user1's VoIP service, user2's VoD service, and user'3 FT service have a SPref values equal to 1, 3, and 2, respectively, the resulting L1 capacity repartition becomes: CL1 = 30 Kbps, CL2 = 600 Kbps, CL3 = 370 Kbps.

CONCLUSION

In this chapter, we have proposed an Autonomic NEtwork Management Archi-tecture (ANEMA). This architecture aims to provide a solution to an autonomic management inspired by the theory of utility and policy based management concepts. It introduces the concept of goal and behavioral policies to comple-ment the existing ones. It defines mechanisms designed to transform high-level human requirements into objectives expressed in terms of NUF. To guide the network in achieving these utilities, ANEMA uses some high-level guidelines (management strategies). These guidelines allow describing the goals related to the NUF. These goals are enforced in the network equipment in the form of behavioral policies and analytical optimization models. Given these policies, the network equipment manage themselves (in particular self-configure and self-optimize) and seek to achieve the target network objectives. To highlight the proposed ANEMA, we have described the guidelines of its deployment in the context of multiservice IP networks and validated the proposed analytical models proposed to resolve the defined network utility functions. We have also implemented a simulator where several scenarios have been tested. The results obtained have shown the approach's capacity to achieve autonomic management based on high-level objectives (operators and users). Moreover, we have implemented a small-scale testbed composed of several autonomic routers and servers implementing the ANEMA components. The objective of this experimentation was to highlight the adaptive behavior of the GAP based on the goals policies and the network situation (users, services, etc.). This experimentation has shown that ANEMA allows autonomic behavior in the network—that is, reflexive behavior (i.e., achieving local decisions) and cooperative behavior (i.e., taking decisions based on other autonomic entities knowledge and situation).

We have defined the CIM extensions needed for implementing our analytical model. Nevertheless, we have used our own API based on the Linux Traffic Control (TC) tool to rapidly test and validate our approach. The next step will be to implement CIM providers for our QoE calculation processes based on the small-footprint CIM broker (SFCB). This way we will use the WBEM technologies to manage CIMOM-enabled autonomic routers. This enhancement will make our architecture compliant to management standards and thus more open and more easily manageable.

This contribution is an intermediate step toward achieving full autonomic management in an operator network as introduced by IBM [35]. Addressing large-scale autonomic networks using ANEMA is mandatory to develop solutions acceptable by operators. In this case, it is important to study the effect of local decisions on the entire network and how these decisions are propagated among the autonomic entities. Hence, it is relevant to study how decisions can impact the global objectives of the autonomic network having limited knowledge. In the context of multioperators and mobile ad hoc networks, it is essential to study the impact of merging different autonomic networks. This could lead to very interesting issues in term of goals conflicts and resolution.

REFERENCES

[1] N. Agoulmine, S. Balasubramaniam, D. Botvich, J. Strassner, E. Lehtihet, and W. Donnelly, "Challenges for autonomic network management," *in First Conference on Modelling Autonomic Communication Environment (MACE'06), Dublin, Ireland*, 2006.

[2] J.O. Kephart and D.M. Chess, "The vision of autonomic computing," *IEEE Computer Society Press*, Vol. 36, No. 1, pp. 41–50, January 2003.

[3] P. Horn, "Autonomic computing: IBM's perspective on the state of information technology," *IBM*, Vol. 15, pp. 1–39, 2001. http://www.research.ibm.com/autonomic/manifesto/.

[4] A.G. Ganek and T.A. Corbi, "The dawning of the autonomic computing era," *IBM System Journal*, Vol. 42, No. 1, pp. 5–18, 2003.

[5] J.O. Kephart and R. Das, "Achieving self-management via utility functions," *IEEE Internet Computing*, Vol. 11, No. 1, pp. 40–48, 2007.

[6] M. Parashar and S. Hariri, "Autonomic computing: An overview," Springer Verlag, Vol. 3566 of LNCS, p. 247259, 2005.

[7] X. Li, H. Kang, P. Harrington, and J. Thomas, "Autonomic and trusted computing paradigms," in *Autonomic and Trusted Computing, Third International Conference, ATC'06, Wuhan, China*, September 3-6, 2006, pp. 143–152.

[8] A. Bandara, E. Lupu, A. Russo, N. Dulay, M. Sloman, et al., "Policy refinement for IP differentiated services quality of service management," *IEEE e-Transactions on Network and Service Management*, Vol. 2, No. 2, 2006.

[9] J. Rubio-Loyola, J. Serrat, M. Charalambides, P. Flegkas, and G. Pavlou, "A methodological Approach toward the refinement problem in policy-based management systems," *Communications Magazine, IEEE*, Vol. 44, No. 10, pp. 60–68, 2006.

[10] A. Lapouchnian, S. Liaskos, J. Mylopoulos, and Y. Yu, "Towards requirements-driven autonomic systems design," *Workshop on the Design and Evolution of Autonomic Application Software (DEAS'05)*, pp. 1–7, 2005.

[11] J. Strassner, D. Raymer, E. Lehtihet, and S. van der Meer, "End-to-end model driven policy based network management," *Proceedings of IEEE International Workshop on Policies for Distributed Systems and Networks (POLICY'06)*, pp. 67–70, 2006.

[12] J. Strassner, *Policy-Based Network Management*, Morgan Kaufman Publishers, ISBN 1-55860-859-1, 2003.

[13] S. van der Meer, W. Donnelly, J. Strassner, B. Jennings, and M.O. Foghlfu, "Emerging principles of autonomic network management," *MACE'06*, Vol. 2, No. 1, pp. 29–48, 2006.

[14] W.E. Walsh, G. Tesauro, J.O. Kephart, and R. Das, "Utility functions in autonomic systems," *Proceedings of the International Conference on Autonomic Computing*, pp. 70–77, June 2004.

[15] J.O. Kephart and W.E. Walsh, "An artificial intelligence perspective on autonomic computing policies," *Proceedings of POLICY'04*, p. 3, 2004.

[16] A. Mas-Colell, M.D. Whinston, and J.R. Green, *Microeconomic Theory*, Oxford University Press, ISBN 0195073401, 1995.

[17] W. Liu and Z. Li, "Application of policies in autonomic computing system based on partitionable server," *Proceedings of the International Conference on Parallel Processing Workshops (ICPPW'07)*, p. 18, 2007.

[18] A. Dey and G. Abowd, "Toward a better understanding of context and context-awareness," GVU Technical Report GIT-GVU-99-22, 1999.

[19] E. Lehtihet, H. Derbel, N. Agoulmine, Y. Ghamri-Doudane, and S. van der Meer, "Initial approach toward self-configuration and self-optimization in IP networks," Management of Multimedia Networks and Services (MMNS'05), Barcelona, Spain, October 24–26, pp. 371–382, 2005.

[20] D.B. Leake, *Case-Based Reasoning: Experiences, Lessons and Future Direction*, Cambridge, MA, MIT Press, 1996.

[21] J. Strassner, N. Agoulmine, and E. Lehtihet, "FOCALE: A novel autonomic networking architecture," Latin American Autonomic Computing Symposium (LAACS'07), Campo Grande, MS, Brazil, 2007.

[22] A. van Moorsel, "Metrics for the Internet Age: Quality of experience and quality of business," Technical Report no: HPL-2001-179, 2001.

[23] D. López, J.E. López, F. González, and A. Sánchez-Macían, "An OWL-S based architecture for self-optimizing multimedia over IP services," *MACE'06*, 2006.

[24] H. Derbel, N. Agoulmine, and M. Salaun, "Service utility optimization model based on user preferences in multiservice IP networks," *International Workshop on Distributed Autonomous Network Management Systems (DANMS'07), Globecom Workshop*, 2007.

[25] S. Shenker, "Fundamental design issues for the future Internet," *IEEE Journal on Selected Areas in Communication*, September 1995.

[26] V. Rakocevic, J. Grifths, and G. Cope, "Linear control for dynamic link sharing in multi-class IP networks," *17th IEE UK Teletraffic Symposium, Dublin, Ireland*, May 2001.

[27] S. Blake, D. Black, M. Carlson, E. Davies, Z. Wang, and W. Weiss, "An architecture for differentiated service," United States, RFC Editor, 1998.

[28] "Linux Traffic Control Web site." [Online]. Available: http://tldp.org/ HOWTO/Trafc-Control-HOWTO/index.html

[29] J. Martin, A. Nilsson, and I. Rhee, "Delay-based congestion avoidance for TCP," *IEEE/ACM Trans. Netw.*, Vol. 11, No. 3, pp. 356–369, 2003.

[30] A. Keller, A. Koppel, and K. Schopmeyer, "Measuring application response times with the CIM metrics model," *j-LECT-NOTES-COMP-SCI*, Vol. 2506, 2002.

[31] V. Agarwal, N. Karnik, and A. Kumar, "An information model for metering and accounting," *Network Operations and Management Symposium, 2004. NOMS 2004. IEEE/IFIP*, Vol. 1, pp. 541–554, 2004.

[32] G. Doyen, "Supervision des réseaux et services pair à pair," PhD Thesis, *INRIA, France*, December 12, 2005.

[33] "Matlab: Mathworks Web site." [Online]. Available: http://www.mathworks.fr/

[34] C. Curescu and S. Nadjm-Tehrani, "Time-aware utility-based QoS optimisation," *Proceedings of the 15th Euromicro Conference on Real-Time Systems (ECRTS'03), IEEE Computer Society,* Silver Spring, MD, p. 83, 2003.

[35] S.R. White, J.E. Hanson, I. Whalley, D.M. Chess, and J.O. Kephart, "An architectural approach to autonomic computing," *Proceedings of the International Conference on Autonomic Computing*, pp. 2–9, 2004.

Chapter 5

Federating Autonomic Network Management Systems for Flexible Control of End-to-End Communications Services

Brendan Jennings, Kevin Chekov Feeney, Rob Brennan, Sasitharan
Balasubramaniam, Dmitri Botvich, and Sven van der Meer
[1] *The work described in this chapter is supported by Science Foundation Ireland, via the
"Federated, Autonomic Management of End-to-end Communications Services (FAME)" strategic
research cluster, grant no. 08/SRC/I1403. For a description of the FAME cluster see
http://www.fame.ie.*

INTRODUCTION

Over the past decade the research community has been actively investigating
how network management systems can be developed to achieve the flexibility
and adaptability required to efficiently operate increasingly complex, heterogeneous, and interconnected networks. In the forefront of these efforts has been
the vision of *autonomic network management* [1] in which a network has the
ability to self-govern its behavior within the constraints of human-specified
business goals. Autonomic network management promises a much more flexible approach to management that seeks to both allow systems to automatically
adapt offered services or resources in response to user or environmental changes
and to reduce operational expenditure for network operators.

While significant progress has been made, work to date has focused on autonomic management in the context of single, well-defined network domains.
Relatively little work has been done on how autonomic management systems
can be *federated* across management domains to provide end-to-end management of communications services. Current network management systems do
perform some coordination across domains, but this is limited in scope to a
small number of predefined information exchanges. In this chapter we argue
that autonomic network management systems should be designed to incorporate
capabilities supporting the negotiation and life-cycle management of federations
involving two or more networks, service providers, and service consumers.

We first briefly review previous work on autonomic network management, focusing on aspects relating to federation. We then discuss our view of federation, introducing a layered federation model. Next we discuss the challenges that must be addressed to achieve federation of networks, their management systems, and the organizations/individuals that operate them. Finally, we discuss an example scenario based on end-to-end management of the delivery of Internet Protocol-based Television (IPTV) content that illustrates the benefits to be gained from adopting a federated autonomic management approach.

AUTONOMIC NETWORK MANAGEMENT: AVOIDING NEW MANAGEMENT SILOS

The term *autonomic computing* was coined by IBM as an analogy of the autonomic nervous system, which maintains homeostasis (which means essentially maintaining the equilibrium of various biological processes) in our bodies without the need for conscious direction [2]. Autonomic computing attempts to manage the operation of individual pieces of IT infrastructure (such as servers in a data center) through introduction of an autonomic manager that implements an autonomic control loop in which the managed element and the environment in which it operates are monitored, collected data is analyzed, and actions are taken if the managed element is deemed to be in an undesirable (suboptimal) state. The IBM *MAPE-K* control loop is made up of *Monitor*, *Analyze*, *Plan*, and *Execute* components, all of which rely on a common *Knowledge* repository.

The autonomic computing vision can be summarized as that of a "self-managing" IT infrastructure in which equipment will have software deployed on it that enables it to self-configure, self-optimize, self-heal, and self-protect. That is, it will exhibit what has come to be known as "self-*" behavior. Clearly this is a powerful vision, and it was therefore natural that the networking community would extend this vision from autonomic management of individual elements to autonomic networking—the collective (self-) management of networks of communicating computing elements. As surveyed by Dobson et al. [3], autonomic networking is a burgeoning research area, integrating results from disciplines ranging from telecommunications network management to artificial intelligence and from biology to sociology.

From the network management perspective, one of the successors to the IBM MAPE-K architecture is the *FOCALE* architecture [1], which seeks to address the particular challenges of managing communications network devices. FOCALE implements two control loops: A *maintenance loop* is used when no anomalies are found (i.e., when either the current system state is equal to the desired state, or when the state of the managed element is moving toward its intended goal); and an *adjustment loop* is used when one or more policy reconfiguration actions must be performed and/or new policies must be codified and deployed. The adjustment loop is supported by a dynamically updateable knowledge base—one that can be modifed to reflect new knowledge at runtime as new knowledge is discovered by reasoning and machine learning components. Furthermore, FOCALE is designed as a distributed architecture in which autonomic control loops can be embedded within autonomic managers, control single or groups of managed elements, and in which semantic models and model-based translation are used to support a common management *lingua franca* that can be used to normalize monitoring information and configuration commands.

Besides FOCALE, other proposed autonomic networking architectures have addressed some aspects of coordination of management activities between networked devices. For example the *Meta-Management System* (MMS) [4] provides robust management-plane communications incorporating self-configuration, self-healing, self-optimization, and self-protection capabilities. Another example is the *Autonomic Network Architecture* (ANA) [5], which aims to be generic and provide flexibility at all levels of an autonomic network by supporting the coexistance of different network styles and communications protocols. While efforts toward development of effective distributed autonomic management systems promise significant benefits, we believe there is a need for a more holistic approach in which communication and coordination of management activities between devices is governed by organizational federations that mandate management systems and autonomic managers to cooperate to achieve business goals.

OUR VIEW OF FEDERATION

We employ *federation* as a general term for describing cross-organizational capability sharing. More specifically, we define a federation as a persistent organizational agreement that enables multiple autonomous entities to share capabilities in a controlled way.

We make a number of observations regarding the implications of this definition. First, a federation brings together *autonomous entities*—organizations or individuals endowed with sovereign decision-making power over the resources they own or control. Hence, there is no single governing authority for the federation. Second, given the lack of a single governing authority, the federation must exist by virtue of the *agreement* of its members. If the federation is to be viable, it is important that its nature, structure, and evolution be agreed upon and that the value delivered to federation members by virtue of their participation be transparent and clear to each of them.

The third observation is that federations exist in order to enable the *controlled sharing of capabilities* between autonomous entities. We use the term *capability* in the widest possible sense. It could range from the availability of a communication channel to the ability to perform device configuration changes. Controlled capability sharing means that federation members are granted (possibly constrained) access to capabilities they would not otherwise possess. Finally, we observe that federations should be *persistent*—which is not meant to imply

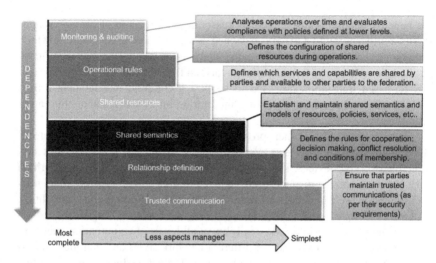

FIGURE 5.1   Layered Federation Model (LFM).

that they are permanent or must last any minimum time period. They should, however, outlive individual transactions or interactions between federation members and will thus have an associated *federation life cycle*.

Following from the above discussion of our understanding of federations, we now introduce the *layered federal model* (LFM), depicted in Figure 5.1. The LFM is intended as a general-purpose high-level conceptual model of the components of a federal agreement. The model is decomposed into layers, with each layer representing particular aspects of a federation agreement on the organizational arrangement. Its main purpose is to serve as a useful model for the decomposition of federal relationships in order to render their definition and maintenance more tractable and transparent. The layers represent the most important aspects of successful persistent, cross-organizational relationships, with the relative positioning of these aspects representing the dependencies between the various elements that constitute a federal agreement. We now briefly review the six layers of the layered federation model.

- Trusted Communication Layer

    A basic requirement for any sort of communication between autonomous entities is a channel with sufficient security measures to satisfy both parties' requirements for the current dialogue. This requires them to agree on communications protocols, security mechanisms, and even applications that can deliver appropriate facilities for identification, authentication, privacy, and accountability. These concerns form the most fundamental layer of our federation model because all higher level agreements and interactions depend on them;

- Federal Relationship Definition Layer

    The relationship definition layer supports the definition and transmission of the basic rules that govern each organization's relationships with other organizations within the federation. This provides a generic layer in which rules concerning membership of the federation and sharing of capabilities (and their revocation) can be negotiated and agreed. For example, to support a peering relationship organizations may create a federation requiring their networks to carry traffic that originated in the networks of other federation members;

- Shared Semantic Layer

  Federations, as we define them, exist to allow autonomous organizations to share capabilities. However, any particular organization will generally have its own addressing mechanisms and semantics for describing the resources that it controls and the capabilities that they support. The shared semantic layer provides mappings between the semantics used internally by each federation member to describe their resources and capabilities to those used by the other parties. This could be achieved by means of a standardized federal semantic language, or it could be achieved by each party mapping directly between their internal semantics and those of the other parties to the federation.

- Shared Capabilities Layer

  Having established sufficiently secure communications, a general re-source-sharing regime and shared semantics with respect to resources, the prerequisites are in place to allow capabilities to be shared. The capability-sharing layer is concerned with enabling members of federations to manage the dynamic set of capabilities they share. This includes providing a means whereby members of the federation can add and remove capabilities from the pool available to other federation members, as well as allowing other parties to discover which capabilities are available for use at any particular time.

- Operational Rule Layer

  The operational rule layer augments the capability-sharing layer by providing federation members with the ability to view and/or change the configuration of resources provided by other federation members. For example, a communications service provider may be granted the ability to configure a home user's set-top box in order to provision an IPTV service. Clearly, the degree to which resource configurations can be viewed and modified needs to be fully specified in the federation agreement;

- Auditing and Monitoring Layer

  In many cases, the lower layers of the federal model can adequately manage their own auditing, reporting, and compliance assurance. However, federal agreements may be formulated in such a way that compliance and verification is only possible through observing aggregate activity over significant periods of time. So, for example, a federal agreement might include a clause that specifies that each member should more or less provide as many useful resources to the federation as they use. Due to random variations, traffic spikes, and so on, such agreements can only be meaningfully checked over significant periods of time. The monitoring and auditing layer is thus the top layer of this federal model. It is responsible for providing members of the federation with detailed monitoring of their compliance and that of counterparties to federal agreements.

FEDERATION OF NETWORKS

As noted earlier, the main focus of the literature on autonomic networking has been on the development of algorithms and processes that allow networks to self-organize, self-optimize, and self-protect. As networks grow in complexity, autonomic capabilities such as this are increasingly seen as a necessity. For example, the anticipated deployment of very large numbers of Femto-cells attached to the mobile access networks will require a significant rethinking of traditional centralized operations and management approaches that are not designed to effectively handle management of very large numbers of network elements. In developing these algorithms and processes, researchers have drawn inspiration from myriad sources, notably the self-management behavior exhibited by biological systems. This has resulted in proposals of varying complexity and with varying responsiveness profiles, many of which have been shown to be suitable for deployment in large-scale network domains. Nevertheless, the question of how self-management algorithms and processes deployed in interconnecting network domains can be coordinated to deliver the best possible end-to-end behavior has received little attention.

Current network domains are predominantly managed on an individual basis, with the aim being to optimize the transfer of information within administrative boundaries. Although cordination across network boundaries does exist, its scope is limited primarily to participation in end-to-end routing protocols, network peering arrangements, and exchange of certain management information (in particular for charging and billing purposes). Furthermore, such coordination is highly static in nature, with changes requiring time-consuming negotiations and coordinated deployment by network operators.

We argue that there are significant benefits to be harnessed through adoption of approaches to coordinating the operation of self-management algorithms and processes across network administrative boundaries. The means of achieving this coordination can be either direct or indirect. In the direct approach, self-management processes would use agreed-upon protocols to directly transfer control information that allows them to synchronize their behavior. In the indirect approach, coordination would be effected by coordination of the parameterization of self-management processes by the management systems associated with the individual networks, with cross-domain interaction being handled by the management systems. Both approaches necessitate the availability of some of the facilities provided by the layered federation model, in particular: the presence of trusted communications across network boundaries, the definition of a federal relationship and some degree of shared semantics between the network operators.

FEDERATION OF MANAGEMENT SYSTEMS

As alluded to above, network management systems typically seek to deliver the best performance from a collection of network elements directly under their control. Even within a single network domain, network elements typically conform to different programming models and are controlled via a myriad of management protocols and interfaces. Management systems themselves rely on system models that, though nominally adhering to industry standards, are in reality very proprietary in nature and are based on divergent semantic models. In this context the task of federating at the management system level is very challenging, but, we believe, necessary to deliver end-to-end autonomic management solutions.

A key requirement for autonomic management systems is the ability to monitor their own environment so that they can react to changes and gather information to populate or specialize their own internal environmental models for planning and prediction. To provide end-to-end autonomic management, it is not sufficient for individual management systems to simply monitor their own domains. Instead, to close the federated, end-to-end autonomic control loop it is necessary to deploy processes that provide service-level monitoring of the operational state of network elements across network boundaries. Such processes would need to provide on-demand and continuous monitoring facilities that can be used to both accurately identify the source of unexpected service degradation and support prevention of potential service degradation. Collected data would need to be shared across different administrative boundaries and processed in a distributed, scalable manner that is conducive to analysis and contextualization using the aid of information or semantic models.

Monitoring data harmonization, as envisaged here, goes far beyond syntactical interworking: It requires mechanisms for mapping and federation of the large volume of low-level data produced by monitoring systems into semantically meaningful information and knowledge that may be discretely consumed by federation members. Any complex system has domain experts who are familiar with how the constituent parts of the system can be and are particularly aware of the end-to-end operating constraints of those constituent parts. Encoding this expertise in a manner that can be utilized by other stakeholders is a difficult task [6], but one we believe can be solved through the application of knowledge representation and engineering techniques. The benefit of enabling such expertise to be encoded, aggregated, and interpreted across diverse domains is that it will allow federation members to monitor other parts of the system, using knowledge gained to inform management decisions made locally. This clearly implies the presence of well-defined federal relationships, sharing of capabilities, and well-formed operational rules and federation monitoring and auditing facilities, as outlined in the layered federation model.

The flipside of end-to-end service level monitoring is how to effect configuration changes in response to sensed changes in a manner that is coordinated across an end-to-end service delivery path. Autonomic management systems seek to analyze monitoring information and use knowledge inferred from it to trigger policy management analysis and decision processes that result in the generation and deployment of sets of device configurations that best align a network's behavior with business goals. However, current network management systems are significantly impaired by their inability to automate the translation of business information to network device configuration commands. Automating the translation of business-level policies (specified in terms of entities such as products, services, and customers) through a number of levels of abstraction into corresponding device instance-level policies (configuration commands specified in terms of entities such as packet marking rules or firewall configuration rules) is hugely challenging, since information needs to be added (and removed) as the policies become more specific in nature. Ensuring the consistency of configurations across a federation is further complicated by the requirement for shared semantics, the need for synchronization, and the presence of appropriate federation operational rules.

FEDERATION OF ORGANIZATIONS AND THEIR CUSTOMERS

To respond to an increasingly deregulated environment, there is a strong impetus for network operators to support more flexible business models. One trend in particular is the proliferation of virtual operators that own little or no traditional infrastructure of their own, so that service delivery is always based on crossing multiple administrative domains and there is never completely in-house service instance handling. Flexibility for providers of carrier networks to deal with multiple virtual providers, perhaps even on a per-service basis, can drive down costs and significantly increase business agility. The benefits for smaller operators are obvious, but even for the big players, support of such flexible infrastructure is becoming a source of competitive advantage as the market for wholesale services continues to grow, driven by the increasing numbers of small operators that use their services. Ultimately, these complex value chains of business relationships also need to be managed themselves. Other trends, including the emergence of prosumers (users who are both producers and consumers of content and services), reinforce the need for operators to explicitly empower their networks to support the creation and management of federations with other network providers, service providers, and their own customers.

To efficiently support shorter service life cycles within dynamic federations of service providers and their customers, autonomic network management systems will need the ability to dynamically negotiate and manage federations with other management systems, while achieving an appropriate balance between satisfying local and federation-wide goals. This will require major extensions to current approaches for expression of business goals, negotiation mechanisms, and distributed security. A particular challenge relates to the difficulty of modeling the differing nature of diverse management approaches. Different priorities will be placed on the value of federation membership by different resources and services, as well as by the managers and/or management systems of those resources and services. This will influence the degree to which actors may wish to exert their management authority at any one time or to delegate it to other actors in the federation.

Increased management flexibility and extensibility can be provided by policy-based management (PBM) and the semantic modeling of management information [1]. PBM is an increasingly popular method for combining flexibility with efficiency in systems and network administration [7]. In PBM systems, administrators encode operational management decisions as rules, which are then mapped into concrete device configurations by the policy system. However, structural abstractions, such as roles and domains, used in many policy rule languages, reflect hierarchical organizational thinking that has been shown to be insufficient for modeling the complex interrelationships between individuals within human organizations [8]. This results in the need for complex tools to manage changes to these policy rule languages, which themselves become obstacles to interoperability and to change. Therefore, handling the authoring and maintenance of a coherent set of policies across a federation of providers presents significant problems for existing PBM systems with regards to policy conflict resolution and collaborative policy consensus forming.

There are significant open issues surrounding how federations are formed, what level of trust is required to support this formation, and what patterns of decentralized authority would best support different forms of federations or value networks. Securing a federation is also challenging, especially as service refinement and innovation may spawn subfederations of providers and customers. In order to ensure that disparate policies are consistently enforced across federations, most existing approaches rely on centralized administrative authority. We believe that noncentralized approaches, such as the distributed authorization language (DAL) [9]—which is designed specifically for decentralized coalition policies and supports dynamic coalition formation, merging, and delegation—will be required to deliver the powerful federation management facilities envisaged for the Operational Rule and Auditing and Monitoring layers of the layered federation model.

Many problems in dynamically federating independent management domains will not be resolved simply by resolving rule (or policy) conflicts because there is no underlying logic for resolving the conflict and, as importantly, verifying that the conflict has been resolved. Rather, such problems will require novel advances in semantic analysis, whereby heterogeneous representations of knowledge describing the state and behavior of managed resources can be harmonized and decisions enforced using unambiguous policy rules. The use of formal semantics, including mappings, in structured knowledge has already been demonstrated for autonomic networking approaches [10, 11]. Using these techniques it is possible to model the resources being managed, the context used in communications services, and the policies used to express governance. For a federated environment, semantic mapping is essential to allow the sharing and common understanding of contracts and policies, which together govern the interactions between different management elements.

We note that traditional semantic mapping approaches, as surveyed in [12], generally assume that the mapping task is performed by a knowledge engineer, whose task is to generate a full static mapping between models that will be used by several applications. This is a constrained and static view of the semantic mapping process that would not meet the requirement for development of a full methodology for integration of disparate knowledge as envisaged in the Shared Semantics layer of our layered federation model. Furthermore, in the context of the LFM Shared Semantics layer, mappings in support of federations will need to be generated to be task-specific and context-sensitive, be able to represent partial knowledge, and need to be tracked, managed, and maintained over time [13].

EXAMPLE SCENARIO: END-TO-END MANAGEMENT OF IPTV SERVICES

We now explore a potential deployment scenario for flexibly supporting federations of networks, their management systems, the organizations that control them, and the users of their services. The scenario focuses on end-to-end delivery of Internet Protocol-based Television (IPTV) content from IPTV service providers' data centers to devices located in users' Home Area Networks (HANs). Deployment of IPTV is a major growth area in the telecommunications market at present, with many network operators seeing IPTV services as the next major driver of revenue. However, large-scale deployment of IPTV presents significant technical challenges. For television services, users have expectations of high levels of Quality of Service (QoS) and Quality of Experience (QoE), however, it is very difficult to ensure that sufficient bandwidth is available across an end-to-end path spanning multiple management domains and heterogeneous device types.

Today, IPTV service provision typically only takes place within a single network operator's domain or between an operator's domain and the domain of a closely associated third-party service provider. A typical current deployment is described by Hu et al. [14], who outline how IPTV services are delivered in Chunghwa Telecom's Taiwanese network. The advantage for operators of controlling all aspects of IPTV service provision is that ownership of the access network to the customer premises gives them the fine-grained control of network capabilities required to provide the necessary QoS/QoE guarantees for real-time media streaming. Furthermore, direct access to configure set-top boxes in customer premises allows them support for enhanced features such as live program guides and interactive, value-added services. Although this provides a workable solution for today's deployments, it is a largely static and tightly coupled solution, necessitated in part by the lack of flexibility of present management systems, particularly with respect to their marginal, if any, support for interdomain capability sharing.

We argue that a more dynamic service provider federation-based approach would enable the customer to pick best of breed third-party service providers and also allow the network operator to partner with a varied array of third parties. To illustrate how this could be achieved, we will now outline how a set of autonomic management techniques could be applied together to provide some of the key functionalities of a federated management solution for IPTV deployments.

Coordinated Self-Management for IPTV Content Distribution

As discussed at the start of this chapter, much of the focus of autonomic network management research has been on the development of self-management algorithms and processes. One of the key challenges is how to coordinate the behavior of self-management processes operating either at different layers within a given network or across network boundaries. In this section we outline two self-management processes: The first is a gradient-based routing process operating in the core network, while the second is a content management process operating at the IPTV service level. The two processes are designed to operate *symbiotically* to provide effective end-to-end delivery of IPTV content to those parts of a network where demand is high. This symbiotic relationship is implicit rather than explicit in that management systems are not required to communicate to ascertain the appropriate parameterizations of processes to allow them to cooperate efficiently. The particular IPTV deployment we consider in this chapter is depicted in Figure 5.2. It includes a single network provider, two independent IPTV service providers, and a number of end-user home area networks (for simplicity, only one is shown in the figure).

The Gradient-Based Routing (GBR) process is a bio-inspired process that, in contrast to existing routing processes, is highly distributed and dynamic, allowing routes to be discovered on a hop-by-hop basis as traffic flows through a network. Central to the process is a gradient solution that mimics the chemotaxis motility process of microorganisms, whereby a microorganism senses the chemical gradient toward a food source and migrates by following the increasing gradient path. In an analogous manner, the GBR process creates a gradient field across the network topology, which is then used to ascertain the best routes between source and destination network nodes. A major strength of the process is that as the load patterns in the network change, so too will the gradient field. This results in efficient use of network bandwidth and increased survivability in the face of failure events. More information on the approach can be found in Balasubramaniam et al. [15].

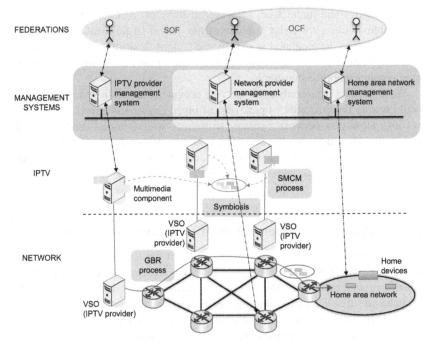

FIGURE 5.2 Federated IPTV Use Case.

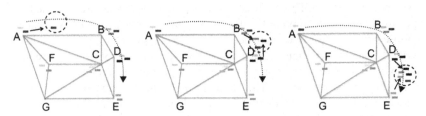

FIGURE 5.3 Example Slime-Mold Inspired Component Composition.

The second process we outline operates at the IPTV service layer; it controls how items of IPTV content are distributed around a set of content servers attached to different network end points and all controlled by a single IPTV service provider. Once again the process is bio-inspired, this time by the behavior of slime molds—organisms that have fungus-like properties and that usually live on dead plant material (see Figure 5.3). A typical property of slime molds is their ability to coordinate among themselves and move as a unit in the event of unfavorable environmental changes, in particular the scarcity of food. An example is Dictyostelium cells, which, when starved, are able to develop into agar that forms fruiting bodies. These fruiting bodies mobilize as a unit and move to a new environment, where, upon arrival, they revert into individual cells.

Our Slime Mold Content Management (SMCM) process applies an analogous process to enhance the management of IPTV content distribution. As the market for IPTV grows, we can expect high levels of volatility in demand for particular pieces of content. Peer-to-peer networks partially address this scenario by segmenting large multimedia content items (such as movies) into short segments that are then distributed around the network. Peer-to-peer techniques are subsequently used to concatenate and stream the content to the end user. This results in significantly improved bandwidth utilization profiles as compared to the traditional approach of having individual flows for each content streaming session. However, such peer-to-peer approaches generally are not aware of the parts of the network experiencing high demand for particular content items or of the prevailing load of different parts of the network. Hence, there is no guarantee that content is distributed optimally from the perspective of the operation of the network.

The SMCM process takes into account the location of user demands and migrates content toward content servers closest to those parts of the networks where those users are attached. As illustrated in Figure 5.3, based on user demands that are diffused and propagated between the nodes, content segments will "absorb" these demands and coordinate and form agar-like units and migrate toward the destination closest to those users' point of attachment. As shown in the figure, as the unit is formed and moves from node, it will pick up segments that are relevant to other segments within the unit. This process will lead to a distributed migration of content items toward parts of the networks having high demand for popular content. This has the effect of lessening the load on the network in comparison to the case where content does not migrate, which makes the task of the GBR process less onerous. Similarly, the operation of the GBR maximizes available bandwidth between content servers and users' points of attachment, which increases the capacity of the IPTV service provider to deliver service to its customers.

As described above, the GBR and SMCM processes implicitly aid each other's management task, providing a form of *mutualistic symbiosis*. This symbiotic relationship could be further enhanced by enabling explicit coordination of the processes. For example, a network provider could provide information relating to the status of the network topology (for example, regions currently experiencing congestion), which the SMCM process could use to divert content migration around congested regions. Similarly, a HAN could provide information relating to user preferences and requirements to the IPTV service provider that could be used to configure how IPTV content items are adapted for consumption by particular users or groups of users. In the next section we explore how network management systems could effect this form of explicit coordination.

Federating Network and IPTV Provider Management Systems

Closer cooperation between management systems is key to realizing flexible federated management. In this context cooperation encompasses the exchange of monitoring and configuration data and possibly the delegation of authority to manage resources. We believe that this form of cooperation is most read-ily achievable where management systems are based on the PBM paradigm. In previous work we [16] addressed the concept of a *Policy Continuum,* in which policies at different levels of abstraction are linked in order to allow high-level goals expressed using business concepts (e.g., customer, service) to be translated into appropriate low-level device configurations (e.g., CLI-based configuration of router traffic shaping and queuing). Implementation of the policy contin-uum enables these constituencies, who understand different concepts and use different terminologies, to manipulate sets of policy representations at a view appropriate to them and to have those view-specific representations mapped to equivalent representations at views appropriate for manipulation by other con-stituencies. If, as illustrated in Figure 5.4, federating management systems each realize a policy continuum, then the governance of the federation can be effected by a set of "federation-level" policies.

Federation-level policies would be formulated to allow federation par-ticipants access to information provided by other participants and/or allow participants to appropriately configure resources in other participants' networks. In previous work [15] we outlined how the creation of a DEN-ng model of the GBR process facilitates its reparameterization by policies. As an example we showed how the GBR could be reparameterized to prioritize optimal load

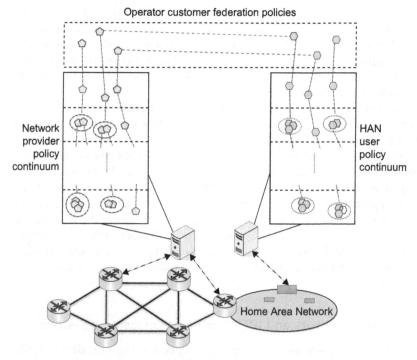

FIGURE 5.4   Operator Customer Federation Policies.

balancing over achieving shortest paths following a link failure event. Similarly, in our IPTV federated management system scenario, a network management system could be allowed to reconfigure GBR parameters in a neighboring network to prioritize end-to-end load balancing when a significant failure occurs in its own network, the objective being to deliver the best possible end-to-end QoS to IPTV service users. Similarly, as discussed above, a network provider management system could have a policy that triggers sharing of network topology and link load information with the management system of an IPTV service provider once loads in parts of the network go beyond a specified threshold. This information would allow the IPTV service provider management system to reparameterize its SMCM process to migrate content away from highly loaded regions of the underlying network.

Realization of management system federation at the policy level will require sophisticated processes for authoring and analyzing policies. Of particular importance is ensuring that deployed policies do not conflict and are consistent with both an organization's own goals and the commitments it makes as part of a federation with other organizations. Therefore, any proposed introduction, withdrawal, or modification of one or more local or federation-level policies must be assessed on this basis. Moreover, as we discuss in the next section, facilities are required to manage the life cycle of the federations themselves.

Interprovider and Provider-User Federations for IPTV Services

As described earlier, the LFM requires the presence of trusted communication and the definition of the federal relationship between participants. There is a substantial body of work on providing functionality of this type (e.g., see [17], which we will not discuss here. Assuming the presence of such facilities, we will now explore the provision of the other aspects of the LFM in the context of our IPTV service delivery scenario.

As illustrated in Figure 5.2, we envisage the deployment of *Federal Relationship Managers (FRMs)* that augment the already deployed management system. For a core network, the FRM may be implemented as a component of the traditional Operations Support Sytem (OSS), while from a HAN the FRM functionality may be integrated as a domain controller in a set-top box. The FRMs act as interconnectors between the management systems of the various organizations participating in a federation, in essence providing the functionality described in the Shared Semantics, Shared Capabilites, and Operational Rules layers of the LFM. They thus enable the establishment of semantic interoperability for the type(s) of integration/interworking and supporting capabilities currently desired by the federation members and enable the identification, publication, discovery, and description of specific capabilities as shared within the federation by individual federation members.

For our IPTV service delivery scenario we address two types of federation: that between the network operator and its customers (OCF) and that between service providers and network operators (SOF). The OCF provides a simple means by which the customer can delegate management authority over the capabilities in the HAN, on a case-by-case basis, to their network provider in the interest of receiving improved customer support, an increased range of services, and ease of use for their HAN. The SOF allows the network operator to make the capabilities shared between the OF available to service providers, including those representing the capabilities that exist in the HANs of individual subscribers. These federations provide the ability to optimize the delivery of IPTV content items by providing a controlled mechanism by which users can allow service providers access to information required to adapt content to that user's device constraints and preferences. Similarly, the network operator and service provider can allow each other access to the information required to maximize the symbiotic relationship between the GBR and SMCM self-management processes.

We view the FRM as an interconnector between existing management systems and existing semantic spaces, rather than a universal model that must be applied across the entire network of relationships. It encapsulates the sum-total of common technical infrastructure that an organization must adopt in order to manage and maintain an arbitrarily complex set of federal relationships. It does not mandate any particular policy language, information model, or management structures or process across a set of federal relationships. In our ongoing work to create FRM prototypes we are utilizing outputs of previous work on an ontology mapping framework [13] and on the *Community Based Policy Management System (CBMPS)* [18]. We now briefly describe these technologies and their application in the context of an FRM.

The CBPMS is a general-purpose policy management framework designed to be policy-language and information-model neutral. Rather than focusing on the specific semantics of policies or resources, it provides a flexible and secure authority management capability. The CBPMS supports decentralized management through delegation of capability authorities. Capability authorities are references to nodes on a resource-authority tree, and this tree is implemented as a service that can be deployed by the owner of any resource that is to be shared. What makes the CBPMS particularly suitable for application in an FRM is that these capability authorities are higher-level constructs than are permissions— the standard unit of most access control and management policy systems. This allows, for example, a telecommunications service provider to grant access to an IPTV service provider to all of their customers (or whatever subset they require) via a single delegation of a capability authority rather than having to specify individual permissions for each user, which is impractical on such a scale. In the CBPMS schema, policies are also associated with capability authorities, which serve as filters on policy searches. When compared to the standard role-based approach to policy management, this yields considerable gains in terms of the size of the policy search space that must be traversed for each policy decision. In large, complex, end-to-end service delivery scenarios, such gains are extremely important, since services such as IPTV have relatively low tolerance for delays in establishing the connections and even lower tolerance for latencies and jitter.

The ontology mapping framework we apply is an extended version of that described in [13] which spans ontology mapping creation, through use and reuse, to evolution and management. The goal of applying this famework within the FRM is to enable the effective and efficient creation and management of mappings to increase understanding of shared capabilities across the federation. For example, in our IPTV scenario this would include the capabilities shared within federations (typically network or service resources) and the federation-level policies used to express governance over the capabilities. Current ontology mapping approaches can be characterized as follows: "knowledge engineers" engage in "one-shot" processes that result in static "one size fits all" mappings, which are then published for indiscriminate use. In contrast, our ontology mapping process is designed to: (a) cope with the diversity of actors involved in managing a federation (i.e., not always specialist knowledge engineers with specialist tools); b) allow for the diversity of ontology mapping execution deployments; (c) enable rich annotation of ontology mappings through meta-data (see below); and (d) enable sustainable and scalable deployment of mappings through dependency modeling.

SUMMARY AND OUTLOOK

Network management systems today federate to provide end-to-end delivery of communications services. However, this federation is achieved through closed or static engineering approaches targeted toward well-defined application scenarios. While these solutions can be very effective, they typically involve significant recurring maintenance costs. In this chapter we argued that this approach will soon be no longer tenable. The continually evolving nature of large networks means that they are ever-changing in terms of the details of individual capability. Hence static federation approaches are fundamentally brittle and expensive in the medium term. In contrast the approach we advocate centers around dynamic creation and management of federations based on negotiating and enabling appropriate, minimal integration between deployed network and service management systems.

In the chapter we outlined a layered federation model that attempts to encapsulate and interrelate the various models, processes, and techniques that will be required to realize such a flexible approach to management system federation. Using an example use case based on end-to-end IPTV service delivery, we discussed the ongoing work of the Ireland-based FAME strategic research cluster [19] in pursuance of this vision. It is hoped that the outcome of this work will go some way toward making flexible federation of future autonomic network management systems a realistic possibility.

REFERENCES

[1] B. Jennings et al., "Towards autonomic management of communications networks," *IEEE Commun. Magazine*, Vol. 45, No. 10, pp. 112–121, October 2007.

[2] J.O. Kephart and D.M. Chess, "The vision of autonomic computing," *Computer*, Vol. 36, No. 1, pp. 41–50, January 2003.

[3] S. Dobson et al., A survey of autonomic communications, *ACM Trans. Auton. and Adapt. Syst.*, Vol. 1, No. 2, pp. 223–259, December 2006.

[4] H. Gogineni, A. Greenberg, D. Maltz, T. Ng, H. Yan, and H. Zhang, "MMS: An autonomic network-layer foundation for network management." *IEEE Journal on Selected Areas in Communications*, Vol. 28, No. 1, pp. 15–27, January 2010.

[5] G. Bouabene, C. Jelger, C. Tschudin, S. Schmid, A. Kelle, and M. May, "The autonomic network architecture," *IEEE Journal on Selected Areas in Communications*, Vol. 28, No. 1, pp. 1–14, January 2010.

[6] J. Strassner, *Policy-Based Network Management: Solution for the Next Generation*, Morgan Kaufmann, 2003.

[7] R. Boutaba, and I. Aib, "Policy-based management: A historical perspective," *J. Netw. Syst. Management*, Vol. 15, No. 4, pp. 447–480, December 2007.

[8] J. Moffett, and M. Sloman, "Policy hierarchies for distributed system management," *IEEE JSAC*, Vol. 11, No. 9, pp. 1404–1414, December 2003.

[9] H. Zhou and S.N. Foley, "A Framework for Establishing Decentralized Secure Coalitions," *Proc. IEEE Computer Security Foundations Workshop*, 2006.

[10] J.E. López de Vergara, V.A. Villagra, and J. Berrocal, "Applying the Web Ontology Language to management information definitions," *IEEE Commun. Magazine*, Vol. 42, No. 7, pp. 68–74, July 2004.

[11] J. Keeney, D. Lewis, D. O'Sullivan, A. Roelens, A. Boran, and R. Richardson, "Runtime semantic interoperability for gathering ontology-based network context," *Proc. IEEE/IFIP Network Operations and Management Symp. (NOMS 2006)*, pp. 55–66, 2006.

[12] N. Noy, "Semantic integration: A survey of ontology-based approaches," *SIGMOD Record*, Vol. 33, No. 4, pp. 65–70, December 2004.

[13] D. O'Sullivan, V. Wade, and D. Lewis, "Understanding as we roam," *IEEE Internet Computing*, Vol. 11, No. 2, pp. 26–33, March/April 2007.

[14] C. Hu, Y Hsu, C. Hong, S. Hsu, Y. Lin, C. Hsu, and T. Fang, "Home network management for IPTV Service Operations—A service provider perspective," Proc. 5th IFIP/IEEE International Workshop on Broadband Convergence Networks (BcN 2010), pp. 1–7, April 2010.

[15] S. Balasubramaniam, D. Botvich, B. Jennings, S. Davy, W. Donnelly, and J. Strassner, "Policy-constrained bio-inspired processes for autonomic route management," *Comp. Netw.*, Vol. 53, No. 10, pp. 1666–1682, July 2009.

[16] S. Davy, B. Jennings, and J. Strassner, "The policy continuum—Policy authoring and conflict analysis," *Comp. Commun.*, Vol. 31, No. 13, pp. 2981–2995, 2008.

[17] S.N. Foley, and H. Zhou, "Authorisation subterfuge by delegation in decentralised Networks," *Proc. 13th Int'l Security Protocols Workshop*, 2005.

[18] K. Feeney, D. Lewis, and V. Wade, "Policy based management for Internet communities," *Proc. 5th IEEE Int'l Workshop on Policies for Distributed Systems and Networks (Policy 2004)*, pp. 23–34, 2004.

[19] FAME Strategic Research Cluster, [Online], Available (last accessed December 4, 2010): http://www.fame.ie

Chapter 6

A Self-Organizing Architecture for Scalable, Adaptive, and Robust Networking

Naoki Wakamiya, Kenji Leibnitz, and Masayuki Murata

INTRODUCTION

Our daily life is enriched—either directly or indirectly—by a considerable number of sensing, computing, controlling, or other information-processing devices, which are placed, distributed, and embedded within our environment. These devices are interconnected and organize themselves as networks to cooperate by sharing and exchanging information and controlling each other. In the future, this trend is expected to lead to an even higher level of ubiquitous and ambient connectivity. To satisfy the wide range of users' requirements and to support our daily life in many aspects, a variety of fixed devices including PCs, servers, home electronic appliances, and information kiosk terminals will be distributed in the environment and connected as networks. Furthermore, mobile devices that are attached to persons and vehicles, as well as small and scattered devices, such as Radio Frequency Identification (RFID) tags and sensors, will require the support of mobility from the network. All those devices generate a great variety of traffic data including voice, video, WWW/file transfer, sensory/control/management data, etc. Their characteristics may also exhibit a large diversity, such as constant/intermittent flow behavior, low/high traffic rate, and small/large volume. Even for a single application, the number, types, locations, and usage patterns of devices, as well as the condition of the communication environment, and traffic characteristics may dynamically change considerably every moment.

In such an environment, a network would often face unexpected or unpredictable user behavior, usage of network, and traffic patterns, which were not anticipated at the time the network was designed or built. As the Internet itself evolved several decades ago from the ARPANET, a testbed network connecting only four nodes, it was never intended for the traffic that is currently being transported over it. With more and more additional protocols being supported, it has been "patched up" to introduce increasingly complex features. In Japan, the rapid and widespread proliferation of broadband access to the Internet, for example, fiber-to-the-home (FTTH), fosters the shift toward an all-IP communication environment, where wireline and wireless phones are assigned an own IP address. However, this step-by-step convergence may result in considerable deterioration of the network performance or, at worst, even total collapse under certain adverse conditions or because of misconfigurations. In February 2009, a single misconfigured Border Gateway Protocol (BGP) router by a Czech ISP caused other routers all over the world to send routing updates, which nearly brought the Internet to its knees. Such an example demonstrates that the conventional network design methodology and architecture, where structures, functionalities, algorithms, and control parameters are designed and dimensioned to achieve their performance based on assumptions on the operating environment and relying on the prepared and preprogrammed detection, avoidance, and recovery of expected failures, no longer seem feasible for new-generation networks. Entirely new approaches are being followed worldwide for a "clean-slate" design [28], avoiding the current dependence on IP, such as FIND [8] and GENI [9] in the United States, Euro-NGI/FGI [4] and FIRE in Europe, and the AKARI Project [1] in Japan.

Taking into account the requirements for new-generation networks, we propose in this chapter a framework for a new network architecture, which has the following properties:

- more scalable to the number of connected nodes and the size of the network
- more adaptive to a wide variety of traffic patterns and their dynamic change
- more robust to expected and unexpected failures independently of their magnitude and duration.

Our fundamental paradigm is to perform the organization and control of the whole network system in a distributed and self-organizing manner. In order to provide interoperability among heterogeneous devices and access media, networks are traditionally constructed as a layered architecture, with the lowest layer being the physical layer and the highest layer being the application layer offering services to the end user. We maintain this structure, but simplify the hierarchy to three layers, with a common layer between physical and service overlay layer that mediates interlayer and intralayer interactions.

The behavior of all entities constituting the network system, that is, that of all nodes, the layers, and the network itself, is entirely self-organized. A node performs medium access control (MAC), scheduling, routing, congestion control, and additional control by using nonlinear functional modules called *self-organization engines*. These engines operate based on local information only that are obtained through observations and exchange among neighboring nodes. Furthermore, nodes are organized and controlled in the network through localized behavior and mutual interactions among them. On a larger scale, networks within the same layer also behave in a self-organizing way and interact with each other directly by exchanging messages and indirectly by influencing the operating environment. In addition to the intralayer interactions, service overlay networks and physical networks interact with each other through the mediation of the common network layer.

In the following sections, we introduce the self-organizing network architecture first, starting with the basic concept and followed by node architecture and components. We conclude this chapter by discussing related work and future issues.

PRINCIPLES OF SELF-ORGANIZATION

One of the key points of the autonomous operation of systems is its feature of *self-organization* of a complex system. While this does not necessarily imply a connection to biologically inspired systems, we will in the following base our focus mostly on self-organizing models inspired by nature and biology since nature provides an abundance of examples of self-organizing systems, as we will address later in this chapter.

Definition of Self-Organization

In essence, a self-organized system can be characterized by having no centralized and global control unit dictating the system behavior. On the contrary, each individual in the system's population acts purely on its own and is not even aware of the objectives of the entire system. It only follows its own set of rules, and the global organization emerges from the sum of each individual's contribution.

Transferred to the context of information networks, this is reflected by the duality among client/server and peer-to-peer (P2P) networks. In P2P networks, each entity (peer) may act simultaneously as client or as server, whereas in a traditional client/server architecture, only a single server (or group of servers) provides the information to the clients. This means that each peer can determine its behavior on its own without relying on the decision of a central server. Obviously, the distributed P2P system is capable of scaling better with an increase in traffic, since the server's processing capacity is limited.

In order to clarify our intended meaning of such systems, we follow the definition of the term *self-organization* as coined by Dressler [6]:

Self-organization is a process in which structure and functionality (pattern) at the higher level of a system emerge solely from numerous interactions among the lower-level components of a system without any external or centralized control. The system's components interact in a local context either by means of direct communication or environmental observations, and, usually without reference to the global pattern.

Self-Organization in Biological Systems

Self-organization occurs in nature or biological systems inherently, since there is no central "leader" that determines how these systems should operate.

Self-organization in biological systems with regard to network applications is discussed in [6] and [23].

In general, four basic principles can be found in the self-organization of biological systems [3]. *Positive feedback*, for example, recruitment or reinforcement, permits the system to develop itself and promotes the creation of structure by acting as an amplier for a desired outcome. On the other hand, *negative feedback* reacts to the influence on the system behavior caused by previous bad adaptations. Furthermore, it contributes to the stability by preventing overshoot control. Another important feature is that biological systems usually do not require a global control unit, but operate entirely distributed and autonomous. Each individual of the system, for example, cells or swarms of insects, acquires, processes and stores its information locally. However, in order that a self-organized structure can be generated, information must be somehow exchanged among the individuals. This is done by *mutual interactions* in the form of either direct interactions (e.g., visual or chemical communication) or indirect interactions among all individuals by influencing the environment (e.g., diffusion of pheromones in slime mold or ant trails). Such indirect interaction through the environment is called *stigmergy*, which is one of the key principles in building self-organizing networks without introducing much overhead involved in direct interactions. Finally, another key characteristic that greatly enhances the stability and robustness of the system is that many decisions are not performed in a deterministic and direct way, but system-inherent *noise* and *fluctuations* drive the system to enable the discovery of new solutions. The following two examples llustrate how self-organized and autonomic behavior can be found in biological systems.

Autonomic Nervous System While organs have certain functionalities, some of which may be essential to the body's operation (e.g., brain, heart), often redundancy can also be found to assist a centralized control in order to provide resiliency. For example, while most neurosensory processing is performed through the *central nervous system* (CNS), which connects the spinal cord with the brain, the internal organs are controlled through the *autonomic nervous system* (ANS) and can operate independently of the CNS.

The ANS's purpose is to control key functions, such has heart and circulation, lungs and respiratory system, kidneys, and so on, and it consists of the *sympathetic* and *parasympathetic* systems, which operate complementary to each other. While the sympathetic ANS is driven mainly by emotional reactions, which are commonly referred to as the "fright, fight-and-flight" response and increases arterial pressure, respiratory activity, and pupil dilation, it prepares the body for intense exertion. On the other hand, the parasympathetic ANS restrains the sympathetic action, calming down the nerves and providing "rest and digest" response [31]. Similarities between the ANS and the Internet dynamics in terms of congestion level have been quantitatively shown in [10].

Ant Colony Optimization *Ant colony optimization* (ACO) is an optimization algorithm that was first proposed to obtain near-optimal solutions of the traveling salesman problem [5]. The algorithm takes inspiration from the

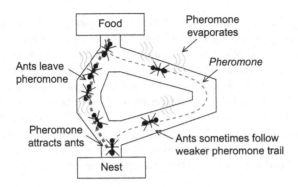

FIGURE 6.1   Ant Colony Optimization as Example Including All Four Basic Self-Organizing Features.

foraging behavior of ants, that is, pheromone trails. Without a centralized control mechanism, ants establish the shortest path between their nest to a food source. The mechanism is explained as follows. Ants come out of the nest and randomly walk around to find food. When an ant finds food, it returns to the nest leaving a small amount of a chemical substance called pheromone. The pheromone left on the ground attracts other wandering ants and guides them to the food. Those guided ants reaching the food also leave the pheromone while carrying food back to the nest, and the pheromone trail is thus reinforced. Since the pheromone evaporates over time, more pheromone is left on a shorter path than on a longer path, and thus a shorter path attracts more ants and has a higher pheromone concentration. However, ants do not act deterministically. Not all ants follow the shortest path with the highest pheromone concentration. Instead, some ants occasionally take longer paths or even a wrong route. Therefore, longer and alternative paths are also kept, which brings the robustness feature to the pheromone trails in ACO. When the primary shortest path is obstructed or lost for some reasons (e.g., water flooding), a spare longer path continues to guide ants to the food, collects pheromones, and becomes the new primary path.

In the ant colony example, all four basic principles of self-organization appear. First, the shortest path attracts more ants and collects more pheromones, which further attracts more and more ants (positive feedback). However, due to evaporation, no path can collect an infinite amount of pheromones (negative feedback). Pheromones left by an ant on the ground attract other ants and influence their behavior (mutual interactions). Finally, ants occasionally take a longer or wrong path (fluctuations); see Figure 6.1.

PROPOSAL OF A SELF-ORGANIZING NETWORK ARCHITECTURE

As the number of nodes and the size of the network increase, a centralized or semidistributed mechanism, in which all nodes are required to have the same global view of the network, as in a table-driven routing, becomes inappropriate. Maintaining global information would lead to considerable maintenance overhead in order to collect and keep up-to-date and consistent information on the whole network system. Therefore, fully distributed and autonomous control mechanisms are required, which enable a node to operate on local information only, which is obtained through observation of its surroundings and through exchange with neighbors, without the need for global information. With such autonomous mechanisms, it also is possible to avoid letting a single and local failure, for example, link disconnection, affect the whole system by, for example, propagating the failure information to update the topology information that all nodes have.

Network Architecture

A conventional adaptation mechanism, whereby the whole system is periodically reoptimized based on status information, puts too much burden on a large-scale network to adapt to frequent changes in the operating environment. Therefore, we need self-adaptive and self-configurable control mechanisms. Each node autonomously adapts the control parameters, behavior, and even algorithms and mechanisms in accordance with the state of the surrounding environment.

Furthermore, a conventional network system acquires its robustness by implementing a variety of detection, avoidance, and recovery mechanisms for failures, errors, abuse, extreme operating conditions, and other critical events. Such design methodology makes a network system complicated, monolithic, and even fragile. Therefore, we need simple and module-based control mechanisms where a node, network, and network system are constituted by autonomous, simple, and interacting functional control modules. When a part of modules halts due to unexpected failure, the remaining modules provide the minimum level of network service and provoke an adaptive behavior of other modules and entities. Consequently, the whole network system adapts to the new environment.

In summary, so that a network system can keep providing network services to users and applications independently of the size of the system and condition of the operating environment, the degree of their diversity and dynamic changes, as well as the scale and duration of failures, it is important to establish a self-organizing network system. We require the following features:

- Each node consists of autonomous and simple control mechanisms.
- Mutual and local interaction among nodes organize a network.
- Interlayer and intralayer interaction among networks organize the whole network system.

The self-organizing network architecture we propose has three layers. These are:

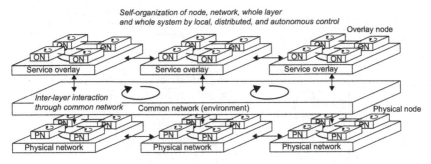

FIGURE 6.2   Self-Organizing Network Architecture.

the *physical network layer* consisting of wireless and wired access networks and optical core networks,

– the *service overlay network* layer consisting of service or application-oriented overlay networks, and

– the *common network layer* mediating between the other two layers.

These layers are self-organized through inter- and intralayer mutual interactions among entities. The architecture is illustrated in Figure 6.2.

Node Architecture

In the self-organizing network architecture, each physical or overlay node consists of communication and *sensing modules, knowledge database modules*, and *self-organization engines* (see Figure 6.3). The communication and sensing module obtains local information through message exchange with neighboring nodes and performs an observation of the environmental condition by probing or sensing, for example. This module also collects status information on the node itself. The obtained information is deposited in the knowledge database to be used by the self-organization engines. A self-organization engine is a basic component for the self-organizing behavior of node. It operates on local information in the knowledge database and reacts to its dynamic changes. By using self-organization engines, a node realizes and performs physical and medium access control, scheduling, routing, congestion control, and other necessary network functionality depending on the layer to which a node belongs. For example, in [33], two self-organization engines are combined, one for adaptive scheduling and routing (network-level control) and the other for adaptive sensing (application-level control) in a wireless sensor network, which shares the same knowledge, that is, observation of sensing targets.

These functionalities range from the physical layer up until the application layer in the seven-layer OSI model. At the time of its introduction, the OSI model was intended to simplify the network architecture and permit the interworking of heterogeneous devices and protocols. However, it has now become

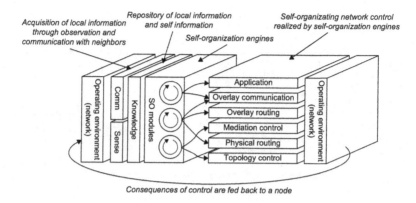

FIGURE 6.3   Node Architecture.

a source of complexity and fragility for the network. For example, to maintain end-to-end communication between mobile nodes, protocols on physical, network, and transport layers need to share the information on the node mobility (e.g., in terms of its velocity) and interact with each other. However, the conventional OSI layer architecture does not allow for such interdependency among layers. In recent years, so-called cross-layer architecture design has been increasingly investigated for its efficiency and effectiveness in tackling new problems such as mobility. The common layer is also introduced to enable cross-layer control, which will be explained in more details later in this chapter.

SELF-ORGANIZATION MODULES

A self-organization engine is a nonlinear functional module that provides a basic networking functionality. It operates on a nonlinear mathematical model in the form of differential equations. Examples of nonlinear models include the pulse-coupled oscillator model [25], the reaction-diffusion model [34], and the attractor selection model [16]. All of these models are derived from the self-organizing behavior of biological systems. Biological systems are inherently fully distributed, autonomous, and self-organizing. As a typical example, it is well known that a group of social insects such as ants, termites, and honey bees often exhibit sophisticated and organized behavior (e.g., generating ant trails, cemetery formation, brood sorting, and division of labor). Such collective intelligence, called *swarm intelligence*, emerges from the mutual and local interaction among simple agents [3].

Pulse-Coupled Oscillator Model

The *pulse-coupled oscillator model* explains the emergence of synchronized behavior in a group of oscillators. Synchronization of biological oscillators can be observed in a group of flashing fireflies, chirping crickets, or pacemaker

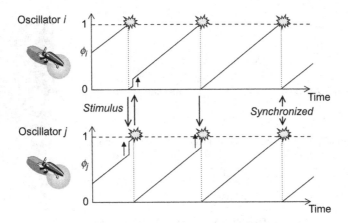

FIGURE 6.4   Pulse-Coupled Oscillator Model.

cells [25]. When it is alone, a firefly periodically flashes based on its biological timer and at its intrinsic frequency. However, when fireflies form a group, a flash of a firefly stimulates nonflashing fireflies, causing stimulated fireflies to advance their timers by a small amount. By repeatedly stimulating each other through flashes, they eventually become synchronized and all fireflies begin to flash at the same time and same frequency (see Figure 6.4).

In a pulse-coupled oscillator model, an oscillator maintains a timer. It fires when the phase of timer ϕ reaches one and then the phase is reset to zero. The dynamics of the phase of the timer is formulated as

$$\frac{d\phi_i}{dt} = \frac{1}{T_i} = \frac{\Delta(\phi_i)}{|\{j|j \in N_i, \phi_j = 1\}|} \sum_{j \in N_i} \delta(1 - \phi_j), \tag{1}$$

In Equation (1), T_i stands for the intrinsic interval of oscillator i's timer and N_i is a set of oscillators coupled with oscillator i. $\Delta(\phi_i)$ is a monotonically increasing nonlinear function that determines the amount of the stimulus. We should note here that the global synchronization, whereby all oscillators flash simultaneously at the same frequency, can be accomplished without all-to-all coupling. As far as a network of oscillators is connected, synchronization eventually emerges from mutual interactions among oscillators. Depending on the parameters and functions, so-called phase-lock conditions, where oscillators flash alternately keeping the constant phase difference, can be accomplished and as a result a traveling wave appears.

A pulse-coupled oscillator model has been applied to a variety of network control such as clock and timer synchronization [30] and scheduling [12, 36].

Reaction-Diffusion Model

The *reaction-diffusion model* describes the emergence of periodic patterns such as spots, stripes, and maze on the surface of animal coat through chemical

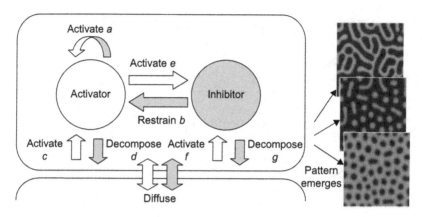

FIGURE 6.5   Reaction-Diffusion Model.

interaction among cells [34]; see Figure 6.5. In the reaction-diffusion model, two hypothetical chemicals, called *morphogens (activator and inhibitor)* are considered. The dynamics of the morphogen concentrations is formulated as

$$\frac{du}{dt} = F(u, v) + D_u \nabla^2 u, \tag{2}$$

$$\frac{dv}{dt} = G(u, v) + D_v \nabla^2 v, \tag{3}$$

where u and v are concentrations of activator and inhibitor, respectively. The first term on the right-hand side of the equations is called the reaction term and expresses the chemical reactions, that is, activation and inhibition, among morphogens. F and G are nonlinear functions stating that the activator activates itself and the inhibitor, whereas the inhibitor restrains the activator. An example of these function is given by

$$F(u, v) = \max\{0, \min\{au - bv + c, M\}\} - du \tag{4}$$

$$G(u, v) = \max\{0, \min\{eu + f, N\}\} - gv \tag{5}$$

which model pattern generation on an emperor angelfish *pomacanthus imperator* [20]. The second term, called the diffusion term, is for interaction among neighboring cells and is expressed by the Laplacian $\nabla^2 u$ and $\nabla^2 v$ and their respective diffusion factors, D_u and D_v. To generate a pattern, the condition $D_u < D_v$,—that is, the speed of diffusion of the inhibitor is faster than that of activator—must be satisfied.

The process of pattern generation can be briefly explained as follows. When the activator concentration occasionally increases due to small perturbations, it activates the activator and inhibitor in the cell. For a faster diffusion rate, the generated inhibitor spreads to neighboring cells and restrains the activator growth at those neighboring cells. On the other hand, a high activator concentration is maintained at the cell and further activates morphogens. Consequently, a spatial heterogeneity in the morphogen concentrations emerges.

Applications of the reaction-diffusion model include autonomous scheduling in a spatial TDMA MAC protocol [7] and optimal cluster formation of wireless sensor network [35].

Attractor Selection Model

The attractor selection model duplicates the nonrule-driven adaptation of *E. coli* cells to dynamically changing nutrient conditions in the environment [16]. A mutant *E. coli* cell has a metabolic network consisting of two mutually inhibitory operons, each of which synthesizes different nutrients, glutamine and tetrahydrofolate. When a cell is in a neutral condition in which both nutrients exist, the concentrations of mRNAs dominating protein production are at a similar level. Once one of the nutrients becomes insufficient, the level of gene expression of an operon for the missing nutrient eventually increases so that a cell can survive in the new environment. A cell can also adapt to the change from one nutrient shortage to the lack of the other by activating the corresponding chemical reaction. However, there is no signal transduction (i.e., embedded rule-based mechanism) from the environment to the metabolic pathway to switch between the expression states of both operons. The dynamics of concentration of mRNAs is formulated as

$$\frac{dx}{dt} = f(x) \cdot \alpha + \eta, \tag{6}$$

where x corresponds to the vector of concentrations of mRNA. $f(x)$ is a function for the chemical reaction in the metabolic network, and α represents the cellular activity such as growth rate and expresses the goodness of the current behavior, that is, gene expression. Finally, η expresses internal and external noise affecting the cell's behavior.

Applying this model to network control, x represents a set of control parameters or control policies. When the current control is appropriate for the environment, activity α, a scalar metric reflecting the goodness of the control, becomes high and the deterministic control function $f(x)$ dominates the system behavior. Once the environmental condition changes and the control becomes inappropriate, activity α decreases and the relative influence of the noise term η becomes dominant. The system then looks for a new appropriate control (i.e., a good attractor) by being driven by random and stochastic control, see Figure 6.6. Eventually the system approaches a new appropriate attractor, leading to an increase in the activity, and the current control is reinforced, driving the system further toward the new attractor. The attractor selection model has been applied to multipath routing in overlay networks [21] and adaptive routing in mobile ad hoc networks [22], where communication is often affected by the unpredictable dynamic behavior of other sessions and mobility of nodes.

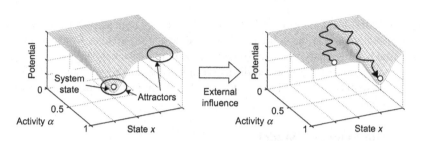

FIGURE 6.6   Attractor Selection Model.

INTER-/INTRA-LAYER INTERACTIONS

The layered definition of the architecture facilitates the interactions among nodes within the same layer and via arbitration of the common middleware layer and also an interaction between service overlay layer and physical network layer. In this section, we will describe how such inter- and intralayer interactions can be performed.

Intralayer Interactions

Nodes operating on self-organization engines directly interact with their neighboring nodes by exchanging messages for stimulation in the pulse-coupled oscillator model and morphogen diffusion in the reaction-diffusion model, for example.

Furthermore, they indirectly interact with each other through environmental changes. Through direct and/or indirect mutual interaction among nodes, a network becomes self-organized.

The autonomous behavior of a single node always has an impact on the environment. For example, when a node changes its communication path from one to another, the old path will get more available communication bandwidth, whereas the new path becomes more highly loaded or congested. In reaction to such environmental changes, other nodes would also change their behavior. This indirect interaction induced by changing the environment, as noted earlier, is called *stigmergy* [3], and it is one of principles of self-organization. In the case of wireless communication, when a node transmits a packet, the wireless channel is occupied by this node for a certain length of time, and it affects the surrounding electromagnetic field. In the self-organizing network architecture, we do not introduce an arbitrator to accommodate the competition for network resources. Instead, nodes adapt their behavior to act symbiotically through mutual interactions. One can find such symbiosis in biological systems, where individuals of different characteristics, properties, and species live together peacefully and comfortably.

Physical networks and service overlay networks also interact with each other in their respective layer. Direct interaction among networks is accomplished by direct message exchanges or mediation of the common network layer. Although mechanisms and interfaces for networks to exchange messages with each other through the common network layer are still under research and development, some approaches such as i3 [32], P4P [39, 40], or TOTA (Tuples On The Air) [24] may be used as a guideline. The i3 architecture is based on a publish-and-subscribe communication mechanism. A node intending to receive packets, such as data or requests, issues an advertisement packet, called *trigger*, consisting of a trigger identifier and a receiver address. A node having corresponding data or intending to receive the service sends a packet consisting of an identifier and data or a message. The packet is sent to the receiver, whose trigger matches the specified identifier. It implies that a source node does not need to know who the receivers or servers are. Networks can exchange messages by letting a designated node register a trigger with an identifier for internetwork communication and sending packets specifying the identifier. P4P is originally an architecture, where an ISP prepares a portal called *iTracker* providing overlay networks with physical network information in order to avoid excessive overhead by overlay networks independently and aggressively probing the physical network condition.

Although the current iTracker only maintains the amount of background traffic and capacity of each link, it can be extended to obtain additional common information useful to the networks. By using such a platform for internetwork interactions, networks are capable of operating in a cooperative manner. Examples of cooperative networks can be found in the literature [18, 26, 37, 35], where networks interact with each other, connect to each other, or even merge to form a single new network, depending on the degree of cooperation and mutual benefit.

Interlayer Interactions

In the OSI layered network model, a network system is divided into seven layers. Each layer operates independently from other layers as far as it conforms to the specifications and satisfies requirements on the SAP (service access point) to upper and lower layers. Such isolation of layers is beneficial in developing and deploying a variety of network protocols, devices, and facilities to deal with heterogeneous characteristics of the operating environment. However, the resulting constructed network system becomes a mixture of independent and individual components and lacks the flexibility to adapt to a dynamic environment.

Recently, especially in the field of wireless networks, the concept of cross-layer design has been attracting many researchers [17]. With regard to the limitation on communication capacity, computational power, and power supply, as well as the instability and unreliability of wireless communication channels,

Higher layer: Control of video coding rate

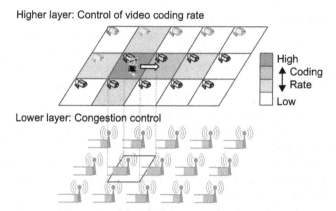

Lower layer: Congestion control

FIGURE 6.7   Reaction-Diffusion Control of a Wireless Camera Sensor Network.

the conventional layered architecture is too complicated and redundant. In a cross-layer architecture, each layer optimizes its behavior, taking into account the information and status of other layers. For example, route establishment based on the wireless link quality expressed by the RSSI (Received Signal Strength Indicator) and the amount of residual energy on nodes incorporate the network layer, the physical layer, and even the management plane.

Now we will present three examples of our work based on interlayer interactions. For details, please refer to corresponding papers. In the first example [14, 41], the same nonlinear mathematical model (i.e., reaction-diffusion model) is used in both the lower and higher layers to control a wireless camera sensor network in a self-organizing manner (Figure 6.7). In the lower layer, the reaction-diffusion model is adopted for congestion control. The reaction term corresponds to a local control in order to balance the occupancy of buffers within a node by adjusting the weight of the WRR (Weighted Round Robin) scheduler. On the other hand, a diffusion term accomplishes global control to balance buffer occupancy among nodes, by adjusting *CWmin* (the minimum of contention window) in relation to the buffer occupancy of neighboring nodes. On the higher layer, the reaction-diffusion model is used to control the amount of video traffic each camera generates so that it will not overcrowd the wireless network while keeping the quality of video capturing of the targeted person high. By exchanging information on the concentration levels of the virtual morphogens among camera sensor nodes and calculating reaction-diffusion equations, they organize a spot pattern centered at the camera capturing the moving targeted person. Then cameras set their video-coding rate in accordance with the morphogen concentrations.

In the second example, the authors in [29] propose an adaptive and energy-efficient data-gathering mechanism for a wireless sensor network by adopting the attractor selection model for clustering and routing (Figure 6.8). In the clustering layer, the activity becomes high when a node with larger residual energy

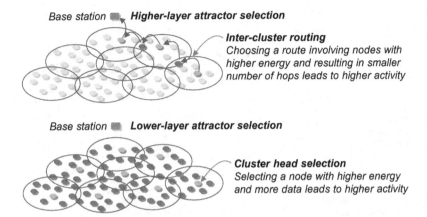

Base station — *Higher-layer attractor selection*

Inter-cluster routing
Choosing a route involving nodes with
higher energy and resulting in smaller
number of hops leads to higher activity

Base station — *Lower-layer attractor selection*

Cluster head selection
Selecting a node with higher energy
and more data leads to higher activity

FIGURE 6.8 Energy-Efficient Data Gathering in Sensor Networks with Layered Attractor Selection.

and larger size of data to transmit is chosen as a cluster head. Then the cluster head chooses one gateway node—that is, a node that simultaneously belongs to the same cluster as the cluster head and another cluster with a cluster head closer to a sink—among the set of gateway nodes by using the attractor selection model. The activity reflects the amount of residual energy of the gateway node and the quality of the path through that gateway node in terms of, for example, the data delivery ratio, delay, and the total amount of data traversing the path. Since the amount of residual energy and the amount of data to transmit dynamically change during operation of a Wireless Sensor Network (WSN), an appropriate cluster formation dynamically changes accordingly. As a result of reclustering, an appropriate path from a cluster head to the sink also changes. The attractor selection model enables a WSN to adapt to such changes and prolongs the network lifetime for as long as possible.

Finally, in [19], a model of layered attractor selection is used to control the topology of a virtual network built on a WDM (wavelength division multiplexing) network in accordance with the behavior of traffic in the higher IP network layer. The attractor selection model consists of two layers—a metabolic pathway network and a gene regulatory network (Figure 6.9). The chemical reaction in the metabolic network is catalyzed by proteins, whose on/off-type of expression levels are controlled by genes and the dynamics are described in the form of Equation (6). The activity in this case corresponds to the cell growth rate, which is determined as an increasing function of concentrations of substrates. In [19], by regarding a WDM network as a gene network, an IP network as a metabolic network, and IP-level performance as growth rate, the WDM network adaptively and dynamically configures lightpaths between IP routers to obtain high IP-level performance.

In the proposed self-organizing network architecture, the common network layer allows entities belonging to different layers to communicate with each

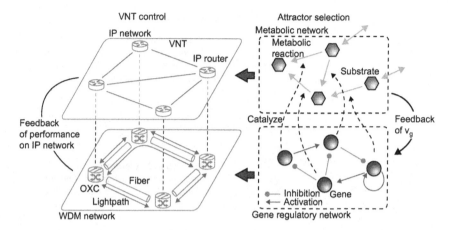

FIGURE 6.9   Attractor Selection Based on Interaction of Gene Regulatory Network and Metabolic Pathway Network.

other in order to exchange and share control information, get feedback from the other layer, and even control the other layer. We should note here that interlayer interaction should be kept loose in order to prevent the introduction of an unnecessarily strong interdependency, which makes the system fragile and leads to unintended consequences.

In a new-generation network constructed on this layered self-organization architecture, small-scale perturbations in the operating environment, such as local congestions, link disconnections, and node failures, are handled by localized and prompt reaction of surrounding nodes. Such localized adaptation could contribute to the prompt response of a network to a localized event or emergency. Network entities around the event do not need to distribute a notification to the whole network or to consult a centralized server for a solution. However, we should note here that especially for an emergency it is not a good idea to rely on the emergence due to its delay in organizing the desired behavior. We would need prepared but simple rules to perform first aid.

A network system constructed on our architecture adapts to large-scale variations, such as the injection of a vast amount of traffic by flooding and spatial and simultaneous failures, by a series of reactions that are induced by mutual interactions among nodes and networks, and that spread over the whole network, layer, and entire network system. From an interlayer control viewpoint, the influence of small-scale physical failures, such as variations in the wireless communication quality and link disconnection, is absorbed in the physical network layer and hidden from the service overlay network layer. On the other hand, in the case of large-scale physical failures, the physical network layer tries to avoid affecting the performance and control of the service overlay network layer, while the service overlay network layer adapts to changes in the physical network configuration. As a result of such cooperative and self-organizing behavior, system-level adaptability, stability, and robustness can be accomplished.

EVALUATION METHODOLOGIES

The main purpose of introducing the self-organizing network architecture is not to improve performance of the network system in terms of conventional measures such as packet delivery ratio, response time, and throughput, but to acquire higher scalability, adaptability, and robustness than ever before. A quantitative evaluation of such *-ties (e.g., scalability, adaptability) and *-ness (e.g., robustness) property is not trivial. In our architecture, since self-organization engines are based on nonlinear mathematical equations, some basic characteristics such as the stability, convergence, and adaptability of each control mechanism can be theoretically analyzed. For example, in the case of the reaction-diffusion model, we can derive the Turing conditions, which the parameters must satisfy in order to generate a pattern and the feasible range of discrete time steps in discretizing equations [13]. However, the collective behavior emerging from interactions among different self-organizing control mechanisms has never been investigated. We are considering incorporating mathematical analysis for the fundamental understanding of nonlinear control and simulation experiments for an in-depth analysis of the emergence of self-organization. For this purpose, we are currently developing a novel simulator whereby the behavior of entities is defined by nonlinear equations and we can observe and investigate their behavior visually. Figure 6.10 shows a snapshot of a simulator, where a virtual

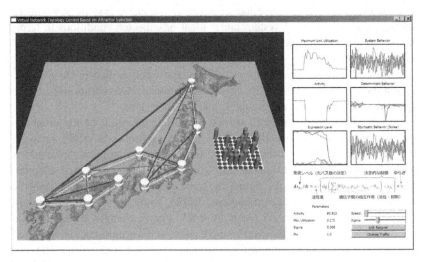

FIGURE 6.10 Screenshot of Simulator Using Attractor Selection for Virtual Network Topology Control.

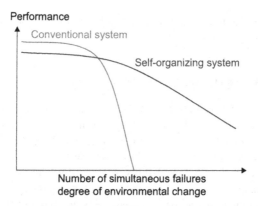

FIGURE 6.11   Comparison between Conventional System and Self-Organizing System.

network topology control mechanism based on attractor selection is evaluated [19]. In this figure, quantitative states of the mechanism can be observed in real time—for example, the maximum link utilization, activity, expression level of genes, deterministic behavior, noise value, and resultant dynamics in the attractor selection-based virtual network topology control described earlier.

Another issue in the evaluation of *-ties and *-ness lies in the definition of the range of parameters and conditions. Do we need to explore an unlimited range of conditions to show the robustness against unexpected failure? The solution to this question still remains to be fully resolved. One of the main goals of the evaluation is to obtain a result as illustrated in Figure 6.11. Because a self-organizing system relies on mutual interactions among entities and there is no centralized control to maximize or optimize the entire system, it would be inferior to a conventional system under ideal or expected conditions in terms of conventional performance measures. However, once the degree of environmental changes including the scale of failures, for example, exceeds a predicted range, a conventional system may suddenly and easily collapse. On the contrary, a self-organizing system can adapt to such unexpected conditions and keep operating, thus making it a sustainable and robust system.

As an example, Figure 6.12 shows a preliminary result of the comparison among schemes of virtual network topology control on a WDM network. "Attractor" in the figure corresponds to the case when an attractor selection model is adopted. On the other hand, "Adaptive" is a heuristic and adaptive scheme based on [11]. In the figure, the performance is evaluated in terms of the ratio of successfully accommodating traffic, while keeping the maximum link utilization lower than 0.5 when the traffic matrix changes at the variance σ^2. As shown in this figure, even in the case of small variations, the obtained performance is the same among the attractor selection-based and the conventional scheme, whereas the former can keep the performance high against high variance.

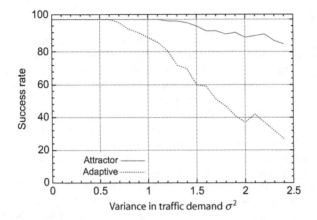

FIGURE 6.12 Comparison between Attractor Selection-Based Control and Heuristic Adaptive Control.

CONCLUSION

In this chapter, we proposed a self-organizing network architecture in which each node, network, layer, and network system is self-organized through intra- and interlayer mutual interactions. The hierarchical architecture of self-* modules can also be found in autonomic computing [15] and the autonomic network [27], but our idea is different by organizing the whole network system following the self-organization principle. For self-organization engines, we use biologically inspired models. Much previous research has been dedicated to bio-inspired and self-organized control [2, 6], but our major difference from the other approaches is that we adopt the biological behavior modeled by non-linear differential equations to enable a mathematical discussion and achieve high adaptability.

Based on our knowledge and experience, we recognize that a self-organizing system is not necessarily optimal and does not always guarantee the best performance. However, we consider it is worth sacrificing performance to some extent to achieve scalability, adaptability, and robustness. We plan to investigate the consequences of intra- and interlayer interactions between different self-organizing behaviors in order to derive a generic design methodology of self-organizing network systems.

ACKNOWLEDGMENTS

The authors would like to thank Yuki Koizumi and Yuki Minami for their contribution. This research was supported in part by the National Institute of Information and Communications Technology (NICT), Japan, and the Global COE (Center of Excellence) Program of the Ministry of Education, Culture, Sports, Science and Technology of Japan.

REFERENCES

[1] AKARI project. New generation network architecture AKARI conceptual design. Report of National Institute of Information and Communications Technology, October 2007.

[2] Sasitharan Balasubramaniam, Dmitri Botvich, William Donnelly, Micheal O Foghlu, and John Strassner, "Biologically inspired self-governance and self-organisation for autonomic networks," *Proceedings of 1st International Conference on Bio-inspired Models of Network, Information and Computing Systems (Bionetics 2006)*, December 2006.

[3] Eric Bonabeau, Marco Dorigo, and Guy Theraulaz, *Swarm Intelligence: From Natural to Artificial Systems*. New York: Oxford University Press, 1999.

[4] European Commission, Network of excellence EuroNGI and EuroFGI design and engineering of the next generation internet. http://eurongi.enst.fr.

[5] Marco Dorigo, Vittorio Maniezzo, and Alberto Colorni, Positive feedback as a search strategy. Technical Report 91-016, Politecnico di Milano, Italy, June 1991.

[6] Falko Dressler, *Self-Organization in Sensor and Actor Networks*, Hoboken, NJ: John Wiley & Sons, November 2007.

[7] Mathilde Durvy and Patrick Thiran, "Reaction-diffusion based transmission patterns for ad hoc networks," *Proceedings of IEEE INFOCOM 2005*, pp. 2195–2205, March 2005.

[8] National Science Foundation, Future Internet design (FIND), http:// find.isi.edu.

[9] National Science Foundation, Global environment for network innovations (GENI), http://www.geni.net.

[10] K. Fukuda, L.A. Nunes Amaral, and H.E. Stanley, "Similarities between communication dynamics in the Internet and the autonomic nervous system," *Europhys. Lett.*, Vol. 62, No. 2, pp. 189–195, April 2003.

[11] A. Gencata and B. Mukherjee, "Virtual-topology adaptation for WDM mesh networks under dynamic traffic," *IEEE/ACM Transactions on Networking*, Vol. 11, pp. 236–247, April 2003.

[12] Yao-Win Hong and Anna Scaglione, "A scalable synchronization protocol for large scale sensor networks and its applications," *IEEE Journal on Selected Area in Communications*, Vol. 23, No. 5, pp. 1085–1099, May 2005.

[13] Kastuya Hyodo, Naoki Wakamiya, Masayuki Murata, Yuki Kubo, and Kentaro Yanagihara, "Experiments and considerations on reaction-diffusion based pattern generation in a wireless sensor network," *Proceedings of 1st IEEE International Workshop: From Theory to Practice in Wireless Sensor Networks (T2PWSN 2007)*, Helsinki, Finland, June 2007.

[14] Ibid.

[15] IBM Corporation, "An architectural blueprint for autonomic computing," Autonomic Computing White Paper, 2004.

[16] Akiko Kashiwagi, Itaru Urabe, Kunihiko Kaneko, and Tetsuya Yomo, "Adaptive response of a gene network to environmental changes by fitness-induced attractor selection," *PLoS ONE*, Vol. 1, No. 1, December 2006.

[17] Vikas Kawadia and Panganamala Kumar, "A cautionary perspective on cross layer design," *IEEE Wireless Communications Magazine*, Vol. 12, No. 1, pp. 3–11, February 2005.

[18] Péter Kersch, Róbert Szabo, and Zoltán Lajos Kis, "Self organizing ambient control space— an ambient network architecture for dynamic network interconnection," *Proceedings of the 1st ACM workshop on Dynamic Interconnection of Networks (DIN'05)*, pp. 17–21, September 2005.

[19] Yuki Koizumi, Takashi Miyamura, Shin'ichi Arakawa, Eiji Oki, Kohei Shiomoto, and Masayuki Murata, "Application of attractor selection to adaptive virtual network topology

control," *Proceedings of 3rd International Conference on Bio-Inspired Models of Network, Information, and Computing Systems (BIONETICS)*, Hyogo, Japan, November 2008.

[20] Shigeru Kondo and Rihito Asai, "A reaction-diffusion wave on the skin of pomacan-thus, a marine angelfish,"*Nature*, 376, pp. 765–768, August 1995.

[21] Kenji Leibnitz, Naoki Wakamiya, and Masayuki Murata, "Biologically-inspired self-adaptive multi-path routing in overlay networks," *Communications of the ACM*, Vol. 49, No. 3, pp. 62–67, March 2006.

[22] Kenji Leibnitz, Naoki Wakamiya, and Masayuki Murata, "A bio-inspired robust routing protocol for mobile ad hoc networks," *Proceedings of 16th International Conference on Computer Communications and Networks (ICCCN 2007)*, pp. 321–326, August 2007.

[23] Kenji Leibnitz, Naoki Wakamiya, and Masayuki Murata., "Biologically inspired networking," in Q. Mahmoud (ed.), *Cognitive Networks: Towards Self-Aware Networks*, Chapter 1, pp. 1–21, Hoboken, NJ: John Wiley & Sons, September 2007.

[24] Marco Mamei, Franco Zambonelli, and Letizia Leonardi, "Programming coordinated motion patterns with the tota middleware," *Euro-Par 2003, International Conference on Parallel and Distributed Computing*, published on LNCS 2790, pp. 1027–1037, Klagenfurt, Austria, 2003. Springer Verlag.

[25] Renato E. Mirollo and Steven H. Strogatz, "Synchronization of pulse-coupled biological oscillators," *Society for Industrial and Applied Mathematics Journal on Applied Mathematics*, Vol. 50, No. 6, pp. 1645–1662, December 1990.

[26] Eli De Poorter, Benoît Latré, Ingrid Moerman, and Piet Demeester, "Symbiotic networks: Towards a new level of cooperation between wireless networks," *International Journal of Wireless Personal Communications*, Vol. 45, No. 4, pp. 479–495, June 2008.

[27] Autonomic Network Architecture Project, http://www.ana-project. org.

[28] Clean Slate Project, http://cleanslate.stanford.edu/index.php.

[29] Ehssan Sakhaee, Naoki Wakamiya, and Masayuki Murata, "Self-adaptability and organization for pervasive computing and sensor network environments using a biologically-inspired approach," *Proceedings of IEEE International Workshop on Modeling, Analysis and Simulation of Sensor Networks (MASSN-08)*, Sydney, Australia, December 2008.

[30] Osvaldo Simeone and Umberto Spagnolini, Distributed time synchronization in wireless sensor networks with coupled discrete-time oscillators, *EURASIP Journal on Wireless Communications and Networking*, 2007, doi:10.1155/2007/57054.

[31] John Stein and Catherine Stoodley, *Neuroscience: An Introduction*, Hoboken, NJ: John Wiley & Sons, June 2006.

[32] Ion Stoica, Daniel Adkins, Shelley Zhuang, Scott Shenker, and Sonesh Surana, "Internet indirection infrastructure," *Proceedings of ACM SIGCOMM*, August 2002.

[33] Yoshiaki Taniguchi, Naoki Wakamiya, Masayuki Murata, and Takashi Fukushima, "An autonomous data gathering scheme adaptive to sensing requirements for industrial environment monitoring," *Proceedings of 2nd IFIP International Conference on New Technologies, Mobility and Security (NTMS 2008)*, number 52–56, Morocco, November 2008.

[34] Alan Turing, "The chemical basis of morphogenesis," *Philosophical Transactions of the Royal Society of London*, B. 237(641), pp. 37–72, August 1952.

[35] Naoki Wakamiya, Katsuya Hyodo, and Masayuki Murata, "Reaction-diffusion based topology self-organization for periodic data gathering in wireless sensor networks," *Proceedings of Second IEEE International Conference on Self-Adaptive and Self-Organizing Systems (SASO 2008)*, October 2008.

[36] Naoki Wakamiya and Masayuki Murata, "Synchronization-based data gathering scheme
      for sensor networks," *IEICE Transactions on Communicatios (Special Issue on Ubiquitous
      Networks)*, E88-B(3), pp. 873–881, March 2005.

[37] Naoki Wakamiya and Masayuki Murata, "Toward overlay network symbiosis," *Proceedings
      of the Fifth International Conference on Peer-to-Peer Computing (P2P 2005)*, pp. 154–155,
      August–September 2005.

[38] Naoki Wakamiya and Masayuki Murata, "Dynamic network formation in ambient infor-
      mation networking," *Proceedings of 1st International Workshop on Sensor Networks and
      Ambient Intelligence (SeNAmI 2008)*, Dunedin, New Zealand, December 2008.

[39] Haiyong Xie, Arvind Krishnamurthy, Avi Silberschatz, and R. Yang, "P4P: Explicit
      communications for cooperative control betweeen P2P and network providers,"
      http://www.dcia.info/documents/P4P~Overview.pdf, 2007.

[40] Haiyong Xie, R. Yang, Arvind Krishnamurthy, Yanbin Liu, and Avi Silberschatz, "P4P:
      Provider portal for applications," in Haiyong Xie (ed.), *Proceedings of ACM SIGCOMM*,
      pp. 351–362, August 2008.

[41] Atsushi Yoshida, Takao Yamaguchi, Naoki Wakamiya, and Masayuki Murata, Z'Proposal of
      a reaction-diffusion based congestion control method for wireless mesh networks," *Proceed-
      ings of the 10th International Conference on Advanced Communication Technology (ICACT
      2008)*, Phoenix Park, Korea, February 2008.

Chapter 7

Autonomics in Radio Access Networks

Mariana Dirani, Zwi Altman, and Mikael Salaun

INTRODUCTION

This chapter focuses on autonomic networking in Radio Access Networks (RANs). There is no unique term for autonomic networking in RANs, although *Self-Organizing Network* (SON) is often used for the family of self-operating or *self-x* functions. Today SON functionalities are considered an integral part of future RANs. They provide a means to simplify network management tasks, to reduce operational costs, and to enhance network performance and profitability. The inclusion of SON functionalities in the 3GPP Long Term Evolution (LTE) standard of beyond 3G networks [1, 2] is illustrative of the new trend in autonomic networking.

Autonomics and Self-Organizing Radio Access Networks

Auto-tuning is a self-optimizing function that aims at dynamically optimizing network and Radio Resource Management (RRM) parameters. Auto-tuning has been among the first SON functionalities studied in RANs. Parameter adaptation allows the network to adapt itself to traffic variations and modifications in its operation conditions, such as the introduction of a new service or a change in the weather and the propagation conditions. Hence a better use of the deployed infrastructure and of the scarce radio resources can be achieved when implementing auto-tuning. Self-optimizing networks have been extensively studied in the last decade for different RAN technologies. More recently, the scope of SON has been enlarged and covers more mechanisms of network operation:

- *Planning*: Network planning includes the assessment of the required infrastructure in greenfield or in densification scenarios in order to meet some coverage and capacity targets. Interfacing network databases of network configuration and of the monitored performance data to a planning tool can simplify the process of network planning. Identifying the needs, the required resources, and their location can be done automatically using automatic planning tools that are interfaced to the network.

- *Deployment*: When deploying network elements such as a new base station, the material should be correctly configured. Self-configuration deals with the automatic installation and parameters' configuration of new network elements.
- *Maintenance*: Self network testing, monitoring, diagnosis, and healing are part of the autonomic maintenance process.

To implement self-operating functions such as self-configuration, self-optimization, or self-healing, self-awareness capability of network elements at different levels is required. Appropriate information should transit between network elements and should be processed. In self-optimizing networks, a network element may need to learn the best action it should perform in a given network state, which is defined by its environment. The entity capable of performing actions that modify the network (i.e., utilized resource, parameters, or RRM algorithm) and that has learning capabilities is called an agent. The network may have many agents operating simultaneously, at each base station, for example. The agents can exchange information and can act in a cooperative manner. The agents' capability of sensing and reacting intelligently to the environment and of performing tasks can be referred to as cognitive mechanisms.

The chapter begins with a brief summary of RRM. Next the concept of SON is presented with the different Self-x functionalities. An overview of SON in different RANs is presented in the following section encompassing Global System for Mobile communications (GSM), Universal Mobile Telecommunications System (UMTS), LTE and the IEEE 1900.x family of standards. This discussion is followed by an investigation of control and learning strategies for auto-tuning, namely, for self-optimizing networks. Two approaches for constructing controllers that orchestrate the auto-tuning process are described. The first is the Fuzzy-Logic Controller (FLC), which implements a rule-based control. The second one uses Reinforcement Learning (RL) techniques that, through a learning process, optimize the control function that maps optimal action to each state of the system. A use case of self-optimizing network in the LTE RAN is presented next using the RL approach with a Dynamic Programming (DP) implementation. A discussion with remarks on SON perspectives in RANs concludes this chapter.

RADIO RESOURCE MANAGEMENT

Certain self-optimizing network functionalities are associated with RRM functions, and their implementation depends on the utilized Radio Access Technology (RAT). A brief overview of basic RRM functions is presented here. The provision of wireless communication services requires managing limited radio resources over the air-interface between wireless users and the network infrastructure. Resources should be efficiently used and shared to guarantee quality of service (QoS), network performance, and profitability.

The way resources are shared among the users differs according to the technology. For example, GSM uses Frequency-Time Division Multiple Access (FTDMA), UMTS–Wideband Code Division Multiple Access (WCDMA), and LTE–Orthogonal Frequency-Division Multiple Access (OFDMA) in the downlink (DL) and Single Carrier-FDMA (SC-FDMA) for the uplink (UL). Some of the important RRM functions are as follows.

- *Handover*: The handover (HO) function is responsible for mobility between the cells of a mobile network. The HO mechanism can be tuned to achieve traffic balancing between the network cells or between cells of different systems. We distinguish between hard-handover and soft-handover according to whether one or more links are maintained during the HO process. One should note that the HO mechanism requires a well-parameterized neighboring list to work properly, that is, to ensure that the mobile is handed over to the neighboring cell offering the best radio conditions and with available resources. When a mobile is in idle (sleeping) mode, its attachment to a base station is governed by the selection/reselection function.
- *Admission control (AC)*: Admission control regulates the entering connections. A new connection is admitted if it can be properly served with the required QoS. AC should avoid overloading situations in which QoS of other users often deteriorates.
- *Load control*: Overload in a cell is typically avoided by the AC or the packet scheduling. However, if the load exceeds a certain target, the load control can take different measures that vary according to the technology such as deny power-up commands received by the mobile (in UMTS), that reduce throughputs of elastic traffic or bit rate of real-time connections, or that hand over mobiles to other carriers, neighboring cells, or other RANs. When no alternative is available, certain connections can be dropped.
- *Power control*: The transmission of a mobile or a base station may interfere with other users in the same cell or in other cells. The former case concerns particularly WCDMA networks, and the latter concerns most systems. Power control can be used to minimize interference. In UMTS, for example, a fast power control allows one to transmit the minimum power required to achieve the required target Signal-to Interference plus Noise Ratio (SINR).
- *Packet scheduling*: The packet scheduler is responsible for allocating data to different users in a manner that maximizes transmission efficiency while maintaining a certain degree of fairness between the cell users. In OFDMA technologies, time-frequency resource blocks are allocated, taking advantage of the channel condition in both time and frequency domains. When Multiple-Input Multiple-Output (MIMO) is implemented, a space–time–frequency allocation is used.
- *Adaptive Modulation and Coding (AMC)*: In EDGE, HSDPA, LTE, and WiMAX, the radio-link data rate is controlled by adjusting the modulation and/or the channel coding rate. The AMC scheme is determined according to the SINR value through the perceived Bloc Error Rate (BLER). For example,

a decrease of SINR will increase the BLER. As a result, a more robust (lower coding rate and less frequency efficient) modulation will be selected.

RRM functions are often distributed in different entities in the network such as the base station, the mobile, a Mobility Management Entity in the LTE RAN, and so on. In the UMTS RAN, many RRM functions are located in the Radio Network Controller (RNC). In the LTE RAN, a flat architecture has been adopted in which different RRM functions are distributed at the enhanced Node B (eNBs). In the context of SON, RRM parameters are auto-tuned to improve utilization of radio resources and ensure a good success rate of RRM procedures.

SELF-ORGANIZING NETWORK

SON functionalities encompass different aspects of network operation. Their implementation can considerably alleviate operational complexity and speed up operational processes. SON aims at improving both network management and optimization. Research on SON has been ongoing for several years. The development of new standards such as the 3GPP LTE has given a renewed impulse to research activities on SON. The important SON functions are:

- Self – configuration
- Self – optimization
- Self – diagnosis
- Self – healing
- Self – protecting

Figure 7.1 presents the RAN architecture with some of the SON functionalities. Self-optimization, healing, or diagnosis functionalities could be implemented within the Operation and Maintenance Centre (OMC) or could be located elsewhere, that is, in a dedicated SON entity. For this reason the corresponding blocks are partially situated in the OMC. The Data and Configuration Management block includes two functions: reporting network configurations to the OMC that can be used, for example, by the planning tool; and data management that includes data collection and processing. Certain data management functions such as filtering are performed outside the OMC. Locating the planning tool inside the OMC aims at interfacing the tool with the network and at updating it with precise network configurations and other performance and QoS data. Furthermore, one can interface different SON functionalities directly with the planning tool.

Two options can be envisaged for implementing self-optimizing functions and, in particular, the auto-tuning of RRM parameters: online and offline auto-tuning. Online auto-tuning operates as a control process with a time resolution varying from seconds to minutes in which RRM parameters are tuned in a control loop. The auto-tuning in this case can be seen as an advanced RRM functionality (see the section Control and Learning Techniques in SON). Offline auto-tuning operates in a time resolution varying from hours to days and can be seen as an optimization process.

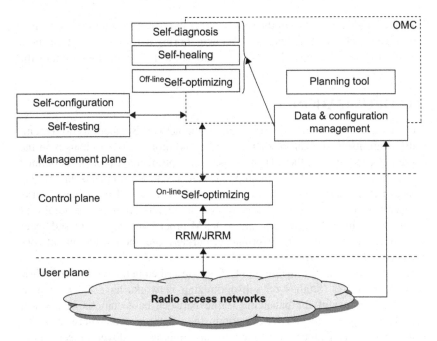

FIGURE 7.1　SON Functionalities within the RAN Architecture.

Self-Configuration

The self-configuration process is defined in the LTE standard as "the process where newly deployed nodes are configured by automatic installation procedures to get the necessary basic configuration for system operation" [1]. Following the deployment of a new base station (or eNB in the LTE technology), different parameters should be configured and functions should be executed such as:

- Hardware and software installation
- Transport network setup (IP addresses, setup QoS parameters and interfaces, etc.)
- Authentication
- Automatic Neighbor Discovery (AND)
- Radio parameter configuration (handover, selection–reselection, power settings, etc.)
- Remote testing

Initial parameter settings can later be improved in the self-optimization process. An example is the AND, or the Automatic Neighbor Relation Function (ANRF), in the LTE standard [1]. The ANRF function is based on the capability of a mobile to send to its serving eNB the Physical Cell Identity (PCI) of the cells it senses. The serving cell can then request from the mobile to send the

Global Cell Identity (GCI) of the sensed eNB, and once it receives this information, it can decide to add this cell to its neighboring list. In a self-optimization process, the neighboring cell list can be updated to follow the evolution of the network [22].

SELF OPTIMIZATION

The term *self-optimizing network* refers to the network's capability to dynamically adapt itself to changes such as traffic variations or other changes in the operating condition. In the LTE standard, self-optimization process is defined as "the process where UE (user equipment) and eNB measurements and performance measurements are used to auto-tune the network. This process works in operational state." Auto-tuning allows full exploitation of the deployed resources, an improved success rate of RRM procedures, and increased spectral efficiency and network performance. Furthermore, the burden on network management is alleviated.

Use cases of self-optimization in LTE are described in the section Overview of SON in RANs. Studies on self-optimizing networks have shown that it is particularly useful for balancing traffic between base stations of a given RAT [10, 11] or between base stations in a heterogeneous network [18]. Traffic balancing can be performed by auto-tuning handover or selection–reselection parameters, and can often bring about important capacity gains. A second approach for load balancing is the Dynamic Spectrum Allocation (DS-Allocation). The idea of DS-Allocation is to share frequency bandwidth between different RANs (capacity pooling). When a RAN needs extra spectrum resources, the DS-Allocation process can locally reallocate these resources from a less loaded RAN in the overlapping area. The DS-Allocation requires first an evaluation of the required spectrum resources according to the traffic demand; then a suitable spectrum subband allocation is performed that maximizes spectral efficiency. A third approach for sharing resources between different RANs is the Dynamic Spectrum Access (DS-Access). The enabling technology to carry out DS-Access is Cognitive Radio technology. The definition of cognitive radio is a "radio (i.e., radio system) that can change its transmitter parameters based on interaction with its environment."

Self-Diagnosis

Network diagnosis refers to the process of identifying the cause of faults in the network. A *cause* could be a hardware failure such as a broken baseband card in a base station or a bad parameter value, that is, transmission power, antenna tilt, or a control parameter such as a RRM parameter. A fault in the network can generate multiple alarms that make it hard to identify its cause. In this case the alarm correlation process may be required to access the root source. A fault may trigger no alarms but result in low performance, poor QoS, and failure of RRM procedures. Often the diagnosis of fault cause requires processing information comprising both alarms and performance indicators. The OMC gathers alarms, counters, and more evolved indicators from the network. A self-diagnosis function can be implemented as an entity within the OMC that can access and process information and perform statistical inference [25].

Self-Healing

Once the cause of a problem has been identified, the self-healing can be applied. If the fault cause is identified with a high degree of confidence (or probability), a correction action can be triggered and be reported to the management entity. If the degree of confidence is not too high, it is proposed to the management entity and is executed only after receiving a manual validation.

Self-Protecting

Security in RANs concerns the ensemble of protection operations such as access control, authentication, and encryption services. The security system monitors the security-relevant contextual input and can reconfigure itself to efficiently respond to dynamic security context. The self-protecting system can modify the cryptographic level, change the strength of the authentication scheme, install system patches to defeat new attacks, and so on.

OVERVIEW OF SON IN RANs

This section presents some of the SON activities for different standards and RAN technologies. The overview does not intend to be exhaustive but rather provides the important trends in SON activities for RANs. First attempts to implement SON functionalities in GSM RAN were reported almost a decade ago. The interest in SON for the UMTS technology increased considerably, but in spite of the important industrial and academic efforts it was not included in the standard. When the Long-Term Evolution (LTE) of UMTS studies were launched by the 3rd Generation Partnership Project (3GPP), the standardization body of GSM, UMTS, and their evolutions [9], the telecom industry was sufficiently mature to implement SON concepts. The SON activity in 3GPP has been supported by both vendors and operators, leading to its inclusion in the release 8 of the 3GPP LTE standard [1]. Parallel to 3GPP, SON concepts and applications are being developed in other standards such as the IEEE 1900 standard.

SON in GSM

In 2001, Ericsson [10] reported an interesting study on the auto-tuning of handover (HO) thresholds to dynamically balance the loads in a hierarchical GSM network. The network comprised macro- and micro-cells in a three-layer layout. The auto-tuning process aimed at dynamically controlling traffic flux from the higher layer (macro-cells) toward lower layers (micro-cells) by adapting the signal strength HO thresholds. It allowed balancing traffic between the cells, leading to a capacity increase, while maintaining QoS. The concept was tested and validated in a field trial.

First contributions on automated diagnosis in GSM have been reported such as [28], which illustrates the potential benefits of automating diagnosis in RANs. This work uses a Bayesian Network model for statistical learning and inference in the automated diagnosis process.

SON in UMTS

Various contributions describing how to dynamically adjust different RRM and system parameters to traffic variations in a UMTS network have been published such as soft HO parameters [11]; downlink load target and maximum link transmitted power [12]; transmitted pilot power [13]; or the uplink (UL) and downlink (DL) Eb/No for packet traffic [14]. See also [15] for a more complete review. Automated diagnosis in UMTS networks has been investigated in [25]. This work extends the Bayesian Network approach to UMTS automated diagnosis.

SON in LTE

3GPP has defined ambitious performance requirements for the Long-Term Evolution (LTE) of the UMTS system in terms of peak and average data rates, spectral efficiency, spectrum flexibility, and so on [19]. The ambitious performance requirements and the need to reduce network management complexity have led to the introduction of SON functionalities in the standard. Different use cases have been studied:

- Coverage and capacity optimization
- Energy savings
- Interference reduction
- Automated configuration of physical cell identity
- Mobility robustness optimization
- Mobility load balancing optimization

Several studies on the Automatic Neighbor Cell Relation List (ANBL) have been carried out [20–22], and this feature has been included in the LTE standard. Interference reduction via Inter-Cell Interference Coordination (ICIC) is of particular importance in OFDMA networks [23]. Different ICIC schemes have been proposed to reduce intercell interference by intelligently allocating frequency resources and/or controlling transmissions or the transmitted power level over certain frequencies. A self-optimizing ICIC scheme is described in detail for the use case described in the section SON Use Case in the LTE Network.

SON in Heterogeneous Networks

Auto-tuning in heterogeneous networks has been studied with the aim of improving network performance by coordinating resource allocation and by improving traffic balancing between the deployed systems. These objectives have been achieved by dynamically adjusting Joint RRM (JRRM) parameters, namely, vertical HO and selection parameters [16–18, 29]. The benefits of dynamically balancing traffic in a heterogeneous RAN context can be considerable.

SON in IEEE 1900 Standard

The IEEE 1900 Standard was established in 2005 with the aim of developing standards that support new technologies and techniques for next-generation radio and advanced spectrum management [3]. An important objective of the standard is to improve the use of the spectrum: manage interference and introduce Dynamic Spectrum Access (DS-Access) and Dynamic Spectrum Allocation (DS-Allocation) mechanisms. The standard also deals with coordinating wireless technologies and includes network management aspects and information sharing. It uses the concept of Cognitive Radio (CR) in the context of DS-Access. S. Haykin gives the following definition to CR [4]:

Cognitive radio is an intelligent wireless communication system that is aware of its surrounding environment and uses the methodology of understanding-by-building to learn from the environment and adapt its internal states to statistical variations in the incoming RF stimuli by making corresponding changes in certain operating parameters (e.g., transmit-power, carrier-frequency, and modulation strategy) in real-time, with two primary objectives in mind:

- highly reliable communications whenever and wherever needed;
- efficient utilization of the radio spectrum.

The IEEE 1900 Standard includes six Working Groups (WGs) that are currently developing the family of standards [3]:

IEEE 1900.1: WG on terminology and concepts for next generation radio systems and spectrum management.
IEEE 1900.2: WG on recommended practice for interference and coexistence analysis.
IEEE 1900.3: WG on recommended practice for conformance evaluation of Software Defined Radio (SDR) Software Modules.
IEEE 1900.4: WG on architectural building blocks enabling network-device distributed decision making for optimized radio resource usage in heterogeneous wireless access networks.
IEEE 1900.5: WG on Policy Language and Policy Architectures for managing cognitive radio for dynamic spectrum access applications.

IEEE 1900.6: WG on spectrum sensing interfaces and data structures for dynamic spectrum access and other advanced radio communication systems.

The concepts developed in IEEE 1900.4 are of particular interest in the context of SON. The standard aims at distributing RRM tasks between the network and the mobiles in a heterogeneous RAN context. A policy-based RRM strategy is proposed in which the network sends to the mobile a policy in the form of rules and constraints. The mobiles then autonomously take the RRM decisions that maximize their utility while complying with the received policy.

CONTROL AND LEARNING TECHNIQUES IN SON

This section describes mathematical approaches for control and learning that can be useful in self-optimizing networks, particularly in auto-tuning RRM parameters. The scheme for auto-tuning is described in Figure 7.2. The RAN can consist of a single or heterogeneous network. Resource allocation and QoS are managed by RRM and JRRM algorithms in the case of distinct or heterogeneous RANs, respectively. The rule based controller is fed by quality indicators

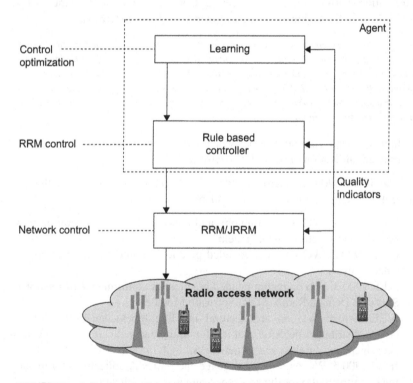

FIGURE 7.2   Self-optimizing Scheme of RRM/JRRM Parameters in a Heterogeneous RAN.

from the network and adjusts or auto-tunes the parameters of the RRM or JRRM algorithms. Hence the auto-tuning is a control process and can be seen as an advanced RRM function. Deriving control rules is often difficult and their design or adjustment is carried out automatically in a learning phase that is an optimization process.

The Agent Concept

The concept of an agent or multi-agents is often used in the context of SON. The controller and the learning block together constitute an agent. The agent receives information from its environment, can act upon it, and has learning capabilities. In next-generation networks such as LTE, eNBs are expected to have SON capabilities, and the entities performing SON functionalities could be agents. The task of RRM parameter auto-tuning is complex since the modification of RRM parameters in one cell may impact performance of its neighboring cells. To achieve high self-optimizing performance, auto-tuning agents in adjacent cells or eNBs need to exchange information, namely, performance indicators such as cell load and QoS indicators. The distributed system with agents exchanging information is a multi-agent system.

Control

The auto-tuning performs a control function that can be modeled as a mapping of states into actions, $\pi : S \rightarrow A$, where S is the state space and A is the action space. The state s, $s \in S$, $s = (s_1, s_2, \ldots, s_n)$, is defined as a vector with n components, which are quality indicators that describe the state of the entity or the subsystem in the controlled network. The state s could have discrete or continuous components. The action a can belong to a discrete set or to a continuous interval.

A simple and elegant framework for designing controllers with continuous states is Fuzzy-Logic Control (FLC) [5], which is a rule based control. The main ideas of FLC are briefly described presently. In classical logic, a predicate is mapped to a Boolean value by a truth function P, $P : X \rightarrow \{0, 1\}$. Fuzzy logic [6] extends the classical logic and allows the mapping of a predicate X belonging to a fuzzy set A to the interval $[0, 1]$, $\mu_A : X \rightarrow [0, 1]$. μ_A is the truth value or the membership function that expresses the degree to which X belongs to A.

In practice, we associate to an observable a number of fuzzy sets with corresponding linguistic labels such as {Small, Medium, High and Very High} or equivalently {S, M, H, VH} (see Figure 7.3). An input blocking rate value of 0.12 is mapped into two fuzzy sets with member function values of 0.8 and 0.2 for the medium (M) and high (H) fuzzy sets, respectively.

Consider the ith element s_i of the state s defined above and its corresponding jth fuzzy set E_{ij}. The predicate "s_i is E_{ij}" has a truth value of $\mu_{ij}(s_i)$. Let m_i be the number of fuzzy sets for the ith component of the state vector. A predicate

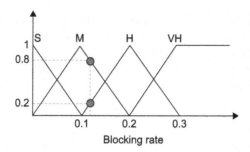

FIGURE 7.3 Fuzzy Sets for the Blocking Rate Variable.

is called an inference rule if it has the form:

$$\text{If } (s_1 \text{ is } E_{1j_1}) \text{ AND } (s_2 \text{ is } E_{2j_2}) \text{ AND } \cdots (s_n \text{ is } E_{nj_n}) \text{ THEN}$$
$$a = o_p; j_i \in \{1, 2, \ldots, m_i\} \tag{1}$$

and is denoted as $R_{j_1 j_2 \ldots j_n}$. The set $\{E_{ij_i}\}$ of a given rule is defined by the modal points (summits) of the corresponding fuzzy sets. It is referred to as the premise part of the rule and is denoted as s_p. To each s_p we can assign a corresponding system state. For sake of simplicity, we refer to s_p as the "state s_p."

Let S_{rules} be the set of rules, p the rule index, P the set of rule indices, and o_p the action of the pth rule. It is assumed here that the controller has a single output, namely, the actions o_p-s of all rules contribute to the modification of the same controlled parameter. An example of a rule of the form (1) can be

IF (Throughput is Low) AND (Interference is High) THEN (Big_Power_Decrease)

The degree of truth α_p of the rule $R_{j_1 j_2 \ldots j_n}$ is written as the product of its membership functions:

$$\alpha_p(s) = \prod_{i=1}^{n} \mu_{ij_i}(s_i) \tag{2}$$

A set of rules of the form (1) with fuzzy sets that cover the domain of variation of the input indicators comprises the controller and can be presented in the form of a table. The action of the controller is given as an aggregation of all the triggered rules and is written as

$$a = \sum_{p \in P} \alpha_p(s) \cdot o_p. \tag{3}$$

An FLC can be designed in a three-step process: Fuzzification (cf. Figure 7.3), Inference (cf. Equation (2)), and Defuzzification (cf. Equation (3)) ([5, 11]). Often, designing the rules and fixing the action values for each rule are complex processes. The design of optimized controllers can be performed automatically via a learning approach and is described in details in the next section.

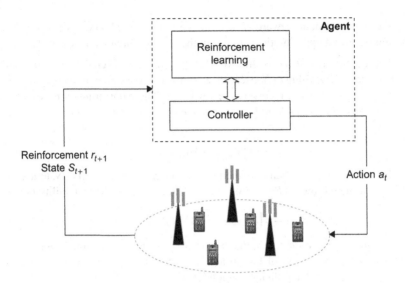

FIGURE 7.4   Reinforcement Learning Scheme.

Learning

This section describes the Reinforcement Learning (RL) approach for designing optimal controllers. According to [7], "RL concerns a family of problems in which an agent evolves while analyzing consequences of its actions, thanks to a simple scalar signal (*the reinforcement*) given out by the environment." A general RL scheme is presented in Figure 7.4. RL is a nonsupervised learning that is guided by the received reinforcement signal. We assume that the investigated RL problems satisfy the Markov property, namely, that the environment (controlled system) response at $t+1$ depends only on the state and action at t. The reinforcement learning task that satisfies the Markov property is called the Markov Decision Process or MDP. When a perfect model of the controlled system is available, the MDP problem can be solved using Dynamic Programming (DP) techniques. When the controlled system model is not known, Temporal Difference (TD) techniques can be applied to solve the MDP. Three algorithms for solving the MDP problem are considered in this section: the Q-Learning (QL) and the Fuzzy Q-Learning (FQL) algorithms for finite and infinite state space, respectively, using a TD implementation; and the Value Iteration Algorithm using DP implementation.

Markov Decision Process

A Markov Decision Process is used to model the interaction between the agent and the controlled environment. The components of a MDP include:

S – the state space, $s \in S$;
A – the set of actions, $a \in A$;

r – the reinforcement (reward) function, $r: S \times A \times S \rightarrow \Re$. $r(s, a, s')$ represents the reward when applying the action a in the state s which leads to the state s'.

The evolution in time of the MDP is presented in Figure 7.4: At a state s_t, the agent (algorithm) chooses an action a_t. The system moves to a new state s_{t+1} and the agent receives a reward r_{t+1}. The MDP transition probability from any state s, $s \in S$, at time t to a state s' given an action a, $a \in A(s)$, is written as

$$P_{ss'}^a = Pr\{s_{t+1} = s' | s_t = s, a_t = a\}. \tag{4}$$

The reward r_{t+1} is obtained when the action a_t is chosen at the system state s_t. The estimated reward $R_{ss'}^a$ is associated with the transition probability (4)

$$R_{ss'}^a = E\{r_{t+1} | s_t = s, a_t = a, s_{t+1} = s'\}. \tag{5}$$

A policy π is defined as a mapping between the state-action couple (s, a) and a probability $\pi(s, a)$ of taking action a in state s. However, we consider here only deterministic policies, namely, $\pi : s \rightarrow a$.

The agent's goal is to maximize the discounted sum of its future rewards, R_t, defined by

$$R_t = \sum_{k=0}^{\infty} \gamma^k r_{t+k+1}, \tag{6}$$

where the coefficient γ is the discount factor that is added to guarantee convergence of the sum in (6) and to give more importance to immediate rewards with respect to future ones. Typically, γ is chosen close to 1, that is, 0.95.

Under a policy π, the value function of a state s, also called the state-value function for policy π, is defined as the expected sum of discounted reward when starting from state s and following the policy π:

$$V^\pi(s) = E_\pi \left\{ \sum_{k=0}^{\infty} \gamma^k r_{t+k+1} \Big| s_t = s \right\}. \tag{7}$$

Similarly, an action-value function is defined for the policy π, $Q^\pi(s, a)$, as the expected return when starting from state s and taking the action a:

$$Q^\pi(s, a) = E_\pi \left\{ \sum_{k=0}^{\infty} \gamma^k r_{t+k+1} \Big| s_t = s, a_t = a \right\}. \tag{8}$$

The optimal policy, denoted by π^*, is defined as the policy maximizing the state-value function, that is,

$$V^{\pi^*}(s) \geq V^\pi(s), \quad \text{for all } s \in S. \tag{9}$$

The optimal state-value function corresponding to the optimal policy is denoted by V^* and the optimal action-value function by Q^*.

The fundamental equation for solving the optimal policy is called the Bellman optimality equation:

$$V^*(s) = \max_{a \in A(s)} \sum_{s'} P_{ss'}^a \left[R_{ss'}^a + \gamma V^*(s') \right] \tag{10}$$

Equation (10) is a recursive equation relating the value function of a given state under the optimal policy to the value functions of the next states under the same policy. This equation expresses the fact that the value of a state under an optimal policy, $V^*(s)$, must be equal to the expected return for the best action from the state s.

The Bellman optimality equation for Q^* is the following:

$$Q^*(s, a) = \sum_{s'} P_{ss'}^a \left[R_{ss'}^a + \gamma \max_{a'} Q^*(s', a') \right] \tag{11}$$

If the probabilities describing the system dynamics are known, the values of $V^*(s)$ and $Q^*(s, a)$ can be recursively calculated using (10) and (11). Conversely, $V^*(s)$ and $Q^*(s, a)$ can be estimated by simulating the time evolution of the system. In the following, two approaches for finding the optimal control are presented.

Dynamic Programming Methods

The algorithms used for solving the optimal control in MDPs are based on estimating the value function at all states. The two classical algorithms used to determine the optimal policy for a MDP are the policy iteration algorithm and the value iteration algorithm, which is presented below [8].

Value Iteration algorithm
1: Initialize v arbitrarily, for example, $V(s) = 0$ for all $s \in S$
2: Repeat
3:     $\Delta \leftarrow 0$
4:         For each $s \in S$:
5:             $v \leftarrow V(s)$
6:             $V(s) \leftarrow \max_{a \in A(s)} \sum_{s'} P_{ss'}^a [R_{ss'}^a + \gamma V^*(s')]$
7:             $\Delta \leftarrow \max(\Delta, |v - V(s)|)$
8: Until $\Delta < \theta$ (a small positive number)
9: Output a deterministic policy, π^*, such that

$$\pi^*(s) = \arg\max_{a} \sum_{s'} P_{ss'}^a \left[R_{ss'}^a + \gamma V(s') \right]$$

The policy iteration algorithm is slightly different from the value iteration algorithm: It performs separately the policy evaluation and the policy improvement.

Q-Learning

When the agent ignores the environment, Temporal Difference methods can be used to solve the MDP problem. The environment is said to be unknown for the agent when the transition probabilities (4) and the reward expectation are unknown.

We denote by $Q_t(s_t, a_t)$ an evaluation of $Q^*(s_t, a_t)$ at time t. $Q_t(s_t, a_t)$ is recursively updated according to the following equation:

$$Q_{t+1}(s_t, a_t) = (1-\eta) Q_t(s_t, a_t) + \eta \left(r_{t+1} + \gamma \max_{a'} Q_t(s_{t+1}, a') \right), \quad (12)$$

where η is a learning coefficient used to avoid premature convergence. The term r_{t+1} is the present reward, and the term $\gamma \max_{a'} Q_t(s_{t+1}, a')$ is the current estimate (at time t) of the sum of discounted future rewards. The value function V_t is defined by:

$$V_t(s) = \max_{b \in A} Q_t(s, b) \quad (13)$$

The Q-function is computed as a table with an entry for each state-action couple. Once the Q-table has been computed, the optimal action for each given state is given by:

$$a = \arg\max_{b \in A} Q_t(s, b) \quad (14)$$

The algorithm is written below.

Q-Learning Algorithm
1: Set $t=0$ and initialize $Q_0(s_0, a)=0$ for all $s_0 \in S$, $a \in A$
2: Observe the initial state s_0
3: Repeat until convergence
4:    $a \leftarrow EEP(s_t, Q_t, S, A)$
5:    Perform action a, leading to a new state s_{t+1}, with a reinforcement r_{t+1}
6:    $Q_{t+1}(s_t, a_t) = Q_t(s_t, a_t) + \eta \{r_{t+1} + \gamma V_t(s_{t+1}) - Q_t(s_t, a_t)\}$
7: Set $s_t = s_{t+1}$, $t=t+1$
8: End repeat

The action choice in Step 4 uses the exploration/exploitation policy (function), $EEP(s_t, Q_t, S, A)$ and is defined as follows:

$$a_t = \begin{cases} \arg\max_{a \in A} Q_t(s_t, a); & \text{exploitation with probability } 1-\epsilon \\ \text{rand}_{a \in A}(a); & \text{exploration with probability } \epsilon \end{cases} \quad (15)$$

The above EEP function implements a policy denoted as the "ε-greedy policy." ε is often chosen as a small probability (i.e., 0.05).

Fuzzy Q-Learning

The Q-Learning assumes a discrete state space and actions. In most practical cases, quality indicators that define the state space are continuous. To handle continuous state space and/or actions, we introduce an interpolation technique based on a fuzzy inference system. The agent learns the optimal policy on a finite set of states $\{s_p\}$ defined by the fuzzy rules $R_{j_1 j_2 \dots j_n}$, namely, the rules' premises (see the earlier section, Control). The action o for each rule can be chosen among a finite set A of competing actions. An action-value quality function q is defined for each state-action couple, (s_p, o) and is stored as a table $q(s_p, o)$, where the subscript p stands for the rule number. Hence the pth rule is defined as:

If s is s_p then $o = o_1$ with $q(s_p, o_1)$

$\qquad$ or $o = o_k$ with $q(s_p, o_k)$

$\qquad$ ....

$\qquad$ or $o = o_K$ with $q(s_p, o_K)$

where $o_1, \dots, o_k, \dots, o_K, o_k \in A$ are possible actions.

The aim of the Fuzzy-Q Learning (FQL) algorithm is to determine the best action for each rule. The Q- and V-functions are defined on continuous states and are computed using interpolation of the q-functions at the states s_p that have a nonzero degree of truth α_p (cf. Equation (2)):

$$Q_t(s, a) = \sum_{p \in P} \alpha_p(s) \, q_t(s_p, o_p) \tag{16}$$

$$V_t(s) = \sum_{p \in P} \alpha_p(s) \max_o q_t(s_p, o) \tag{17}$$

The summation is performed over all the rules. The update equation for the q-function is equivalent to (12):

$$q_{t+1}(s_p, o_p) = q_t(s_p, o_p) + \alpha_p(s_t)\, \eta \left\{ r_{t+1} + \gamma V_t(s_{t+1}) - Q_t(s_t, a) \right\} \tag{18}$$

where s_t is the state currently visited and s_p is the state defined by the p-th rule. The algorithm is written below.

SON USE CASE IN LTE NETWORK: INTERCELL INTERFERENCE COORDINATION (ICIC)

We start this section with a brief description of resource allocation and interference management in the LTE system which serves as an introduction to the use case on intercell interference coordination (ICIC).

The LTE network uses the OFDM transmission technique, which offers the possibility for users to be multiplexed in both frequency and time domains.

Fuzzy Q-Learning Algorithm

1: Set $t=0$ and initialize $q(s_p, o) = 0$ for all $p \in P$, $o \in A$
2: Observe the initial state s_0 and calculate $\alpha_p(s_0)$ for all $p \in P$ using (2)
3: Repeat until convergence
4:     Use EEP policy to determine the action o_p
        with a probability $1 - \epsilon\epsilon$, for each activated rule (with $\alpha_p(s_t) > 0$),
        select an action o_p

$$o_p = \arg\max_{o \in A} q(s_p, o)$$

        with a probability ϵ, for each activated rule (with $\alpha_p(s_t) > 0$),
        select an action o_p

$$o_p = \text{rand}_{o \in A}\{o\}$$

5: Calculate the inferred action a_t (eq. (3))
6: Calculate $Q(s_t, a_t)$ (eq. (10))
7: Execute a_t and observe new state s_{t+1} and reinforcement r_{t+1}
8: Calculate $\alpha_p(s_{t+1})$ for all rules $p \in P$ (eq. (2))
9: Calculate $V(s_{t+1})$ (eq. (17))
10: Update $q(s_p, o_p)$ for each excited rule s_p and action o_p (eq. (18))
11: $t = t + 1$
12: End repeat

Resources are allocated to users with a time resolution of 1 ms. The basic resource unit allocated in the scheduling process is the resource block comprising two slots of 0.5 ms. The nominal resource block has a bandwidth of 180 KHz. Figure 7.5 shows the structure of a physical resource block that corresponds to the grouping of 12 consecutive subcarriers of 15 KHz over a time duration of 0.5 ms. The number of available resource blocks in the LTE system is not specified and can range from 6 to more than 100 depending on the total available bandwidth. The useful symbol time of an OFDM symbol is the inverse of the subcarrier bandwidth ($1/15$ KHz $= 66.7$ μs). Taking into account the cyclic prefix (CP) length, a slot consists of six or seven OFDM symbols (see Figure 7.5), corresponding to a normal and extended cyclic prefix, respectively. Hence, a resource block is a set of 12×7 or 12×6 resource elements depending on the CP length. The possible bandwidth allocations for LTE are 1.25, 2.5, 5, 10, 15 and 20 MHz, and the LTE system can operate in frequency bands ranging from 450 MHz up to at least 2.6 GHz.

Interference Management in LTE system

OFDMA transmission offers orthogonality between resources allocated within each cell. Hence, such systems do not suffer from intracell interference. However, signal quality can be strongly degraded due to intercell interference. To

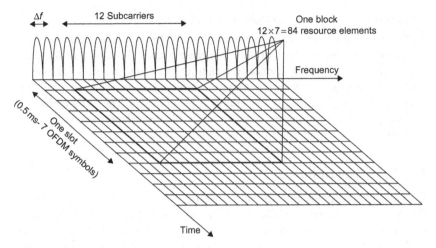

FIGURE 7.5   Downlink Resource Block in the LTE System (normal CP).

improve resource utilization and QoS of cell-edge users, different techniques for interference mitigation have been proposed base on:

- Packet scheduling
- Frequency reuse
- Interference coordination

Packet Scheduling

The scheduling strategy in the LTE system is not specified in the 3GPP standard. Each network equipment manufacturer has the freedom to implement its own scheduling algorithms. A scheduler takes advantage of the channel selectivity in the time and/or frequency domain to schedule the users with the best or relatively best conditions. This suggests a trade-off between opportunistic behavior and acceptable end user QoS.

Frequency Reuse

Frequency reuse schemes have been proposed to improve spectral efficiency and signal quality. The different schemes provide different trade-offs between resource utilization and QoS. The classical reuse-3 scheme proposed for GSM systems offers a protection against intercell interference. However, only a third of the spectral resources are used within each cell. In the reuse-1 scheme in which all the resources are used in every cell, interference at the cell edge may be critical. Fractional reuse schemes are a special case where the reuse-1 is applied to users with good quality (close to the station) and reuse-3 (or higher reuse factors) – to users with poorer quality (close to the cell edge).

Intercell Interference Coordination (ICIC)

ICIC consists in implementing intelligent schemes that allocate radio resources. Static and dynamic scenarios of ICIC can be envisaged. In a dynamic scenario, for example, the ICIC dynamically adapts relevant transmission parameters such as the transmit powers according to the information collected by a dedicated SON entity or directly exchanged between eNBs. The SON entity can impose power restrictions on some resource blocks, adapt antenna tilts and azimuths, or restrict the usage of some resource blocks.

Use Case Description

Consider two cooperating eNBs implementing an ICIC strategy. Figure 7.6(a) illustrates the cooperating eNBs and two interfering ones (colored white) without control. Figure 7.6(b) shows that the bandwidth allocated to the inner zone is shared with that of edge zones of its neighboring cell. The proposed ICIC scheme consists in dynamically adapting the transmitted power to users in the inner zone according to the interference level caused to the neighbouring edge users.

The following classification is used: Users having a ratio between the received Broadcast Channel power and the sum of all Broadcast Channel powers of interfering cells less than a fixed threshold are classified as edge users. The remaining ones are classified as inner users. The two classes are denoted by *e* and *i* for *edge* and *inner*, respectively. It is noted that the classification of users can depend on their geographical location but also on the quality of the received signal that is affected by other radio propagation factors such as shadowing.

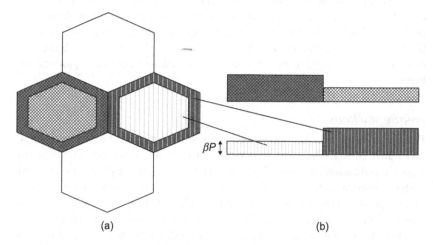

(a)   (b)

FIGURE 7.6   (a) Four Interfering Cells Including Two with ICIC Control and (b) Bandwidth Allocation and Power Restrictions for the Controlled Cells.

A MDP Modeling

Consider a LTE network with random arrivals of users' sessions transferring data files. The arrival process in the system is assumed to be a Poisson process with a mean value of λ_j in the cell j. Such a modeling allows study of performance gain in systems where neighboring cells may experience different and unbalanced traffic. Given the mean throughput experienced by the user, the sojourn time for transferring a data file with exponentially distributed size is also exponentially distributed. The departure rate is dictated by the mean throughput experienced by the users of each cell and class. The average throughput is dictated by the signal statistics and is averaged over all users within the same class. The signal statistics depend on the total power used to transmit the signal by the serving eNB and the experienced interference from neighbouring cells.

MDP States

The states describing the MDP process are defined as the number of users within each class for the two controlled eNBs. A simple admission control limiting the number of admitted users to a predefined value is adopted. The state space is determined by the maximum admissible number of users.

MDP Actions and Policy

The action at a given system state is the power restriction applied to the users of the inner zone of the cell (the β coefficient in Figure 7.6).

MDP Events

Two types of events are considered in this system: (1) The arrival of a new user to the system and (2) the departure of a user from the system after service completion. For sake of simplicity, we assume that the mobiles are still, having no mobility.

The arrival rate of users within a cell j and class $c(c=i,e)$, λ_{jc}, is independent of the actions taken by the system. However, the departure rates are impacted by the power restrictions performed in the cells. The departure rate for a class c in cell j when a total bandwidth BW_{jc} is allocated to users within this class and when users achieve a mean spectral efficiency $\overline{SE}_{jc}$ is given by

$$\mu_{jc} = \frac{BW_{jc}\overline{SE}_{jc}}{E[F]}, \tag{18}$$

$E[F]$ being the mean file size. Decreasing β, namely, the power allocation for the cell center users will decrease their spectral efficiency and, from (18), the departure rate.

MDP Rewards

The control process seeks to maximize a long-term average or discounted reward. The reward can be a performance measure of the system and can be enforced to achieve some fairness or trade-off between conflicting goals. All fairness approaches (system optimization, max-min and proportional fairness) are special cases of the generalized fairness ([26, 27]). When the optimization objective is related to the throughput *th* in the system, the utility (reward) can be expressed as

$$\overline{Th} = \sum_k \frac{Th_k^{1-\alpha}}{1-\alpha}, \tag{19}$$

k being the mobile index. The allocation is globally optimal for $\alpha = 0$, is proportionally fair for $\alpha \to 1$, optimizes the harmonic mean for $\alpha \to 2$, and provides a max-min fairness for $\alpha \to \infty$.

In the present use case, the actions are taken in a distributed manner by each controller. The reward r_j seen by each controller is a global reward taking into account the utility seen by the entire system. We use the Value Iteration Algorithm (see the section Q-Learning) to solve the optimization problem associated with this MDP.

Simulation Results

The simulation parameters are described presently. A total system bandwidth of 4 MHz is considered. The path loss follows the Okumura-Hata model, $PL_{dB} = A + B \times \log_{10}(d_{Km})$, with $A = -128$ and $B = 3.76$. Shadowing is log-normally distributed, and its standard deviation is taken as 6 dB. The noise power spectral density is taken as -173 dBm/Hz. User bit rates associated with the achieved SINR values correspond to the Adaptive Modulation and Coding scheme adopted. Quality tables give the mapping between SINR ranges and their corresponding bit rates. The Poisson arrival process is assumed, and resources are randomly allocated to users within each class of the cell. A data session remains in the system until the complete transfer of the considered file. An average file size of 700 Kbits is assumed. Figure 7.7 compares the mean transfer time of the two cooperating cells for the system with- and without control. One can see significant reduction ranging between 1 and 27% for the mean file transfer time.

Figure 7.8 illustrates the improvement in blocking probabilities. For a fixed target of 5% for the call block rate, one can see the increase of system capacity of about 20%. The improvement in fairness at the session level comes at the expense of a reduction in the system throughput of about 14% as illustrated in Figure 7.9.

The 3GPP has specified ambitious QoS requirements for users within the cell edge. This use shows that to achieve fairness in the session level and to satisfy QoS of edge users, one has to seek optimal trade-offs between cell-edge and global performance.

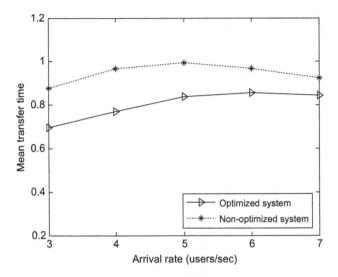

FIGURE 7.7   Mean File Transfer Time in the System.

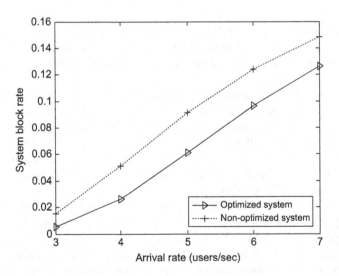

FIGURE 7.8   Total System Call Block Rate.

CONCLUSIONS

The general trend for NGN is to provide better and more attractive services, with growing bit-rates to support high-quality multimedia applications. The market dynamics will continue to put pressure to reduce the cost of the communi-cation services. To face the challenges of NGN, RANs will have to provide much higher spectral efficiencies with lower operation costs. In this context,

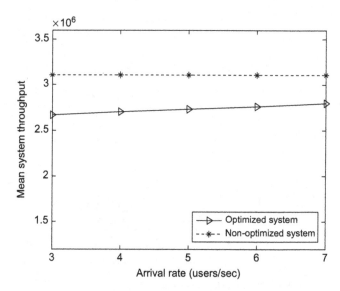

FIGURE 7.9   Mean System Throughput.

radio systems with SON capabilities are currently being developed. Such systems will considerably simplify and reduce costs of management tasks and will improve utilization of the deployed radio resources in heterogeneous RAN landscape.

The capability of learning is a central component of SON mechanisms. Different learning approaches can be envisaged to perform autonomic functions, depending on the nature of the task. Statistical learning is well adapted to self-diagnosis and self-healing tasks, and RL has shown to be effective for self-optimization tasks. The use case presented in this chapter has shown the benefit of SON for dynamic resource allocation in the context of interference coordination in a LTE network.

The area of autonomic networking or SON in RANs is with no doubt in its early stage and will be the topic of growing R&D activity. Among the challenges introduced by SON are making robust algorithms, particularly for self-optimizing and self-healing, that could be considered as reliable by network operators. SON functionalities intervene in different levels and entities in the network in both the management and control planes. In this context, a global and coherent view of autonomic-networking encompassing the different self-x functionalities and their possible interactions should be studied.

REFERENCES

[1]   3GPP TS 36.300, "Evolved Universal Terrestrial Radio Access (E-UTRA) and Evolved Universal Terrestrial Radio Access Network (E-UTRAN), Overall description," Stage 2 (Release8), March 2008.

[2]   3GPP TR 36.902, "Evolved Universal Terrestrial Radio Access Network (E-UTRAN); Self configuration and self-optimization network use cases and solutions," (Release 8), February 2008.

[3]   IEEE Standards Coordinating Committee 41 (Dynamic Spectrum Access Networks), IEEE SA, http://www.ieeep1900.org.

[4]   S. Haykin, "Cognitive radio: Brain-empowered wireless communications," *IEEE JSAC*, February 2005.

[5]   M. Sugeno, *Industrial Applications of Fuzzy Control*, Elsevier Science New Yori, 1985.

[6]   L.A. Zadeh, "Fuzzy sets," *Information and Control*, Vol. 8, pp. 338–353, 1965.

[7]   P.Y. Glorennec, "Reinforcement learning: An overview," *Proc. of the ESIT 2000 conference*, Aachen, Germany, September 2000.

[8]   R.S. Sutton and A.G. Barto, *Reinforcement Learning: An Introduction*, MIT Press, Cambridge, MA, 1998.

[9]   3rd Generation Partnership Project (3GPP), http://www.3gpp.org/.

[10]   P. Magnusson and J. Oom, "An Architecture for self-tuning cellular systems," *Proc. of the 2001 IEEE/IFIP Inter. Symp. on Integrated Network Management*, pp. 231–245, 2001.

[11]   Z. Altman, H. Dubreil, R. Nasri, O. Ben Amor, J.M. Picard, V. Diascorn, and M. Clerc, "Auto-tuning of RRM parameters in UMTS networks," Chapter 16 in *Understanding UMTS Radio Network Modelling, Planning and Automated Optimisation: Theory and Practice*, Eds. M. Nawrocki, M. Dohler, and H. Aghvami, Wiley, Hoboken, NJ, 2006.

[12]   A. Höglund and K. Valkealahti, "Quality-based tuning of cell downlink load target and link power maxima in WCDMA," *56th IEEE Vehicular Technology Conference 2002-Fall*, September 24–28, 2002.

[13]   K. Valkealahti, A. Höglund, J. Parkkinen, and A. Flanagan, "WCDMA common pilot power control with cost function minimization," *56th IEEE Vehicular Technology Conference 2002–Fall*, September 24–28, 2002.

[14]   A. Hämäläinen, K. Valkealahti, A. Höglund, and J. Laakso, "Auto-tuning of service-specific requirement of received EbNo in WCDLA," *56th IEEE Vehicular Technology Conference 2002-Fall*, September 24–28, 2002.

[15]   A. Höglund and K. Valkealahti, "Automated optimization of key WCDMA parameters," *Wireless Communications and Mobile Computing*, Vol. 5, pp. 257–271, 2005.

[16]   A. Tolli, P. Hakalin, and H. Holma, "Performance Evaluation of Common Radio Resource Management (CRRM)," *Proc. of IEEE Inter. Conf. on Communications*, Vol. 5, pp. 3429–3433, 2002.

[17]   A. Tolli and P. Hakalin, "*Adaptive Load Balancing between multiple cell layers,*" in *56th IEEE VTC 2002-Fall*, Vol. 5, pp. 3429–3433, 2002.

[18]   R. Nasri, A. Samhat, and Z. Altman, "A new approach of UMTS-WLAN load balancing; algorithm and its dynamic optimization," *1st IEEE WoWMoM Workshop on Autonomic Wireless Access 2007* (IWAS07), Helsinki, Finland, June 2007.

[19]   3GPP TR 25.913, V7.1.0, "Requirements for Evolved UTRA (E-UTRA) and Evolved UTRAN (E-UTRAN)," (Release 7), September 2005.

[20]   3GPP TSG RAN WG3, R3-071494, "Automatic neighbour cell configuration," August 2007.

[21]   3GPP TSG RAN WG3, R3-071819, "On automatic neighbour relation configuration," October 2007.

[22]   F. Kylvaja, Mikko G. Alford, J. Li, and J. Pradas, "An Automatic Procedure for Neighbor Cell List Definition in Cellular Networks," *1st IEEE WoWMoM Workshop on Autonomic Wireless Access 2007* (IWAS07), Helsinki, Finland, June 2007.

[23]   IST Winner II project, Interference avoidance concept, Deliverable D4.7.2, June 2007.

[24] R. Nasri and Z. Altman, "Handover adaptation for dynamic load balancing in 3GPP long term evolution systems," *5th International Conf. on Advanced in Mobile Computing & Multimedia (MoMM2007)*, Jakarta, Indonesia, December 3–5, 2007.

[25] R. Khanafer et al., "Automated diagnosis for UMTS networks using Bayesian Network Approach," *IEEE Trans. on Vehicular Technology*, July 2008.

[26] T. Bonald, L. Massoulié, A. Proutière, and J. Virtamo, "Queueing analysis of max-min fairness, proportional fairness and balanced fairness," Queueing Systems, Vol. 53, No. 1–2, pp. 65–84(20), June 2006.

[27] C. Touati, E. Altman, and J. Galtier, "Generalized Nash bargaining solution for bandwidth allocation," *Computer Networks*, Vol. 50, No. 17, pp. 3242–3263, December 2006.

[28] R. Barco, V. Wille, and L. Diez, "System for automatic diagnosis in cellular networks based on performance indicators," *European Trans. Telecommunications*, Vol. 16, No. 5, pp. 399–409, October 2005.

[29] D. Kumar, E. Altman, and J.M. Kelif, "Globally optimal user-network association in an 802.11 WLAN & 3G UMTS hybrid cell," *Proc. of 20th International Teletraffic Congress (ITC-20)*, Ottawa, Canada, June 2007.

Chapter 8

Chronus: A Spatiotemporal Macroprogramming Language for Autonomic Wireless Sensor Networks

Hiroshi Wada, Pruet Boonma, and Junichi Suzuki

This chapter considers autonomic wireless sensor networks (WSNs) to detect and monitor spatiotemporally dynamic events, which dynamically scatter along spatiotemporal dimensions, such as oil spills, chemical/gas dispersions, and toxic contaminant spreads. Each WSN application is expected to autonomously detect these events and collect sensor data from individual sensor nodes according to a given spatiotemporal resolution. For this type of autonomic WSN, this chapter proposes a new programming paradigm, *spatiotemporal macroprogramming*. This paradigm is designed to reduce the complexity of programming event detection and data collection in autonomic WSNs by (1) specifying them from a global network viewpoint as a whole rather than a viewpoint of sensor nodes as individuals and (2) making applications behave autonomously to satisfy the spatiotemporal resolutions for event detection and data collection. The proposed programming language, Chronus, treats space and time as first-class programming primitives and combines them as a spacetime continuum. A spacetime is a three-dimensional object that consists of two spatial dimensions and a time playing the role of the third dimension. Chronus allows application developers to program event detection and data collection to spacetime, and abstracts away low-level details in WSNs. The notion of spacetime provides an integrated abstraction for seamlessly expressing event detection and data collection as well as consistently specifying data collection for both the past and future in arbitrary spatiotemporal resolutions. This chapter describes Chronus's design, implementation, runtime environment, and performance implications.

Autonomic Network Management Principles. DOI: 10.1016/B978-0-12-382190-4-00008-5

INTRODUCTION

Wireless sensor networks (WSNs) are considered key enabler to enhance the quality of monitoring and early warning in various domains such as environmental monitoring, emergency response, and homeland security [1–3]. This chapter considers autonomic wireless sensor networks (WSNs) to detect and monitor spatiotemporally dynamic events, which dynamically scatter along spatiotemporal dimensions, such as oil spills, chemical/gas dispersions, and toxic contaminant spreads. With autonomic WSNs, each application is expected to autonomously detect these events and collect sensor data from individual sensor nodes according to a given spatiotemporal resolution. Its goal is to provide human operators with useful information to guide their activities to detect and respond to spatiotemporally dynamic events. For example, upon detecting an event, a WSN application may increase spatial and temporal sensing resolutions to keep track of the event. In order to understand the nature of an event, another WSN application may collect past sensor data to seek any previous foretastes that have led to the current event.

This chapter proposes a new programming paradigm, called *spatiotemporal macroprogramming*, which is designed to aid spatiotemporal event detection and data collection with WSNs. This paradigm is designed to reduce the complexity of programming event detection and data collection by (1) specifying them from a global (or macro) network viewpoint as a whole rather than a micro viewpoint of sensor nodes as individuals and (2) making applications behave autonomously to satisfy given spatiotemporal resolutions for event detection and data collection. The proposed programming language, Chronus, treats space and time as first-class programming primitives. Space and time are combined as a *spacetime* continuum. A spacetime is a three-dimensional object that consists of two spatial dimensions and a time playing the role of the third dimension. Chronus allows developers to program event detection and data collection *to spacetime* and abstracts away the low-level details in WSNs, such as how many nodes are deployed, how nodes are connected and synchronized, and how packets are routed across nodes. The notion of spacetime provides an integrated abstraction to seamlessly express event detection and data collection for both the past and future in arbitrary spatiotemporal resolutions.

In Chronus, a macroprogram specifies an application's global behavior. It is transformed or mapped to per-node microprograms. Chronus is customizable to alter the default mapping between macroprograms and microprograms and to tailor microprograms. It allows developers to flexibly tune the performance and resource consumption of their applications by customizing algorithmic details in event detection and data collection.

This chapter is organized as follows: an overview of a motivating application that Chronus is currently implemented and evaluated for; a description of how Chronus is designed and how it is used to program spatiotemporal event detection and data collection; how the Chronus runtime environment is implemented and how it interprets macroprograms to map them to microprograms; how Chronus allows for customizing the mapping from macro-programs to microprograms; imulation results to characterize the performance and resource consumption of applications built with Chronus; and a conclusion with some discussion on related work and future work.

A MOTIVATING APPLICATION: OIL SPILL DETECTION AND MONITORING

Chronus is designed generically enough to operate in a variety of dynamic spatiotemporal environments; however, it currently targets coastal oil spill detection and monitoring. Oil spills occur frequently[1] and have enormous impacts on maritime and on-land businesses, nearby residents, and the environment. Oil spills can occur due to, for example, broken equipment of a vessel and coastal oil station, illegal dumping, or terrorism. Spilled oil may spread, change the direction of movement, and split into multiple chunks. Some chunks may burn, and others may evaporate and generate toxic fumes. Using an in-situ network of sensor nodes such as fluorometers,[2] light scattering sensors,[3] surface roughness sensors,[4] salinity sensors,[5] and water temperature sensors,[6] Chronus aids developing WSN applications that detect and monitor oil spills. This chapter assumes that a WSN consists of battery-operated sensor nodes and several base stations. Each node is packaged in a sealed waterproof container and attached to a fixed buoy.

In-situ WSNs are expected to provide real-time sensor data to human operators so that they can efficiently dispatch first responders to contain spilled oil in the right place at the right time and avoid secondary disasters by directing nearby ships away from spilled oil, alerting nearby facilities or evacuating people from nearby beaches [6, 7, 11]. In-situ WSNs can deliver more accurate information (sensor data) to operators than visual observation from the air or coast. Moreover, in-situ WSNs are more operational and less expensive than radar observation with aircrafts or satellites [11]. In-situ WSNs can operate during nighttime and poor weather, which degrade the quality of airborne and satellite observation.

---

[1] The U.S. Coast Guard reports that 50 oil spills occurred on U.S. shores in 2004 [4], and the Associated Press reported that, on average, the U.S. Navy caused an oil spill every two days from fiscal year 1990 to 1997 [5].

[2] Fluorescence is a strong indication of the presence of dissolved and/or emulsified polycyclic aromatic hydrocarbons of oils. Aromatic compounds absorb ultraviolet light, become electronically excited, and fluoresce [6]. Different types of oil yield different fluorescent intensities [7].

[3] High intensity of light that is scattered by water indicates high concentration of emulsified oil droplets in the water [8, 9].

[4] Oil films locally damp sea-surface roughness and give dark signatures, so-called slicks [6].

[5] Water salinity influences whether oil floats or sinks. Oil floats more readily in salt water. It also affects the effectiveness of dispersants [10].

[6] Water temperature impacts how fast oil spreads; it spreads faster in warmer water than in cold water [10].

CHRONUS MACROPROGRAMMING LANGUAGE

Chronus addresses the following requirements for macroprogramming.

- **Conciseness.** The conciseness of programs increases the ease of writing and understanding. This can improve the productivity of application development. Chronus is required to facilitate concise programming.
- **Extensibility.** Extensibility allows application developers to introduce their own (i.e., user-defined) programming elements such as operators and functions in order to meet the needs of their applications. Chronus requires extensibility for developers to define their own operators used in a wide variety of applications.
- **Seamless integration of event detection and data collection.** Existing macroprogramming languages consider event detection and data collection largely in isolation. However, WSN applications often require both of them when they are designed to detect and monitor spatiotemporally dynamic events. Chronus is required to provide a single set of programming elements to seamlessly implement both event detection and data collection.
- **Complex event detection.** Traditional event detection applications often consider a single anomaly in sensor data as an event. However, in spatiotemporal detection and monitoring, it is important to consider more complex events, each of which consists of multiple anomalies. Chronus is required to express complex events in event detection.
- **Customizability in mapping macroprograms to microprograms.** Different WSN applications have different requirements such as minimizing false-positive sensor data, latency of event detection, and power consumption. Therefore, the default mapping between macroprograms and microprograms may not be suitable for a wide range of applications. Chronus is required to be able to customize the mapping according to application requirements.

Chronus is designed as an extension to Ruby.[7] Ruby is an object-oriented language that supports dynamic typing. The notion of objects combines program states and functions, and modularizes the dependencies between states and functions. This simplifies programs and improves their readability [12]. Dynamic typing makes programs concise and readable by omitting type declarations and type casts [13, 14]. This allows application developers to focus on their macroprogramming logic without considering type-related housekeeping operations.

In general, a programming language can substantially improve its expressiveness and ease of use by supporting domain-specific concepts inherently in its syntax and semantics [15]. To this end, Chronus is defined as an embedded domain-specific language (DSL) of Ruby. Ruby accepts embedded DSLs, which extend Ruby's constructs with particular domain-specific constructs instead of

---

[7]www.ruby-lang.org

building their own parsers or interpreters [16]. With this mechanism, Chronus reuses Ruby's syntax/semantics and introduces new keywords and primitives specific to spatiotemporal event detection and data collection such as time, space, spacetime, and spatiotemporal resolutions.

Chronus leverages *closures*, which Ruby supports to modularize a code block as an object (similarly to an anonymous method). It uses a closure for defining an event detection and a corresponding handler to respond to the event as well as defining a data collection and a corresponding handler to process collected data. With closures, Chronus can concisely associate handlers with event detection and data collection specifications.

Chronus also employs *process objects*, which Ruby uses to encapsulate code blocks. It allows application developers to define their own (i.e., user-defined) operators as process objects.

Chronus simultaneously supports both data collection and event detection with WSNs. The notion of spacetime allows application developers to seamlessly specify data collection for the past and event detection in the future. It enables WSN applications to perform event detection and event collection at the same time.

Chronus allows developers to define three types of complex events for event detection: *sequence, any,* and *all* events. Each complex event is defined with a set of events. A sequence event fires when a set of events occurs over time in a chronological order. An any and all event fire when one of or all of defined events occur(s).

Chronus leverages the notion of attribute-oriented programming to customize the default mapping from macroprograms to microprograms. Attribute-oriented programming is a program-marking technique to associate program elements with extra semantics [17, 18]. In Chronus, attributes are defined as special types of comments in macroprograms. Application developers can mark macroprogram elements with attributes to indicate that the elements are associated with certain application-specific requirements in, for example, packet rouging. The default macroprogram-to-microprogram mapping can be customized by changing attributes that mark program elements in macroprograms.

Data Collection with Chronus

In Chronus, a data collection is executed one time or periodically to collect data from a WSN.

A data collection pairs a *data query* and a corresponding data handler to process obtained data. Listing 8.1 shows an example macroprogram that specifies several queries. Figure 8.1 visualizes this program.

A spacetime is created with the class Spacetime at Line 4. In Chronus, a class is instantiated with the new() class method. This spacetime (sp) is defined as a polygonal prism consisting of a triangular space (s) and a time period of an hour (p). Chronus supports the concepts of absolute time and relative time.

A relative time can be denoted as a number annotated with its unit (Week, Day, Hr, Min or Sec) (Line 3).

Listing 8.1: An Example Macro-Program for Data Collection

```
1  points = [ Point.new( 10, 10 ), Point.new( 100, 100 ),
     Point.new( 80, 30 ) ]
2  s = Polygon.new( points )
3  p = RelativePeriod.new( NOW, Hr -1 )
4  sp = Spacetime.new( s, p )
5
6  s1 = sp.get_space_at( Min -30, Sec 20, 60 )
7  avg_value = s1.get_data( 'f-spectrum', AVG, Min 3 ) {
8    | data_type, value, space, time |
9    # the body of a data handler comes here. }
10
11 spaces = sp.get_spaces_every( Min 5, Sec 10, 80 )
12 max_values = spaces.collect { |space|
13   space.get_data( 'f-spectrum', MAX, Min 2 ){
14     | data_type, value, space, time |
15     # data handler
16     if value > avg_value then ...  }}
17
18 name = 'f-spectrum'
19 event_spaces =
20   spaces.select{|s| s.get_data(name, STDEV, Min 5)<=10)}.
     select{|s|
21       s.get_data(name, AVG, Min 5) - spaces.prev_of(s).
         get_data(name, AVG, Min 5)>20)} }
```

get_space_at() is a method that the Spacetime class has (Table 8.1). It is called on a spacetime to obtain a snapshot space at a given time in a certain spatial resolution. In Line 6, a snapshot space, s1, contains sensor data available

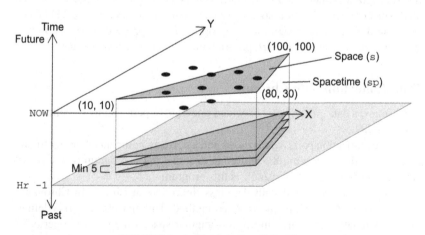

FIGURE 8.1 An Example Data Collection.

TABLE 8.1 Key Methods in Chronus

Method	Description
Spacetime::get_space_at()	Returns a snapshot space at a given time
Spacetime::get_spaces_every()	Returns a set of snapshot spaces
Spacetime +/- spacetime()	Returns a union/difference of two spacetimes
Space::get_data()	Executes a data query
Space::get_node()	Returns a particular node in a space
Space::get_nodes()	Returns a set of nodes in a space
Space::get_border()	Returns a set of nodes that are located at the border of a space
Space::divide()	Divides a space into sub spaces
Space +/- Space()	Returns a union/difference of two spaces
List::collect()	Returns a value from each element in a list
List::select()	Returns a subset of elements in a list

TABLE 8.2 Data Aggregation Operators in Chronus

Operator	Description
COUNT	Returns the number of collected data
MAX	Returns the maximum value among collected data
MIN	Returns the minimum value among collected data
SUM	Returns the summation of collected data
AVG	Returns the average of collected data
STDEV	Returns the standard deviation of collected data
VAR	Returns the variance of collected data

on at least 60% of nodes (the third parameter) in the space at 30 minutes before (the first parameter) with a 20 second time band (the second parameter).

get_data() is used to specify a data query. It is called on a space to collect sensor data available on the space (the first parameter) and process the collected data with a given operator, (the second parameter). Chronus provides a set of data aggregation operators shown in Table 8.2. In Line 7, get_data() obtains the average of fluorescence spectrum ('f-spectrum') data in the space s1. The third parameter of get_data() specifies the tolerable delay (i.e., deadline) to collect and process data (three minutes in this example).

get_data() can accept a data handler as a closure that takes four parameters: the type of collected data, the value of collected data, the space where the data is collected, and the time when the data is collected. In this example, a code block from Line 8 to 9 is a closure, and its parameters contain a string "f-spectrum," the average fluorescence spectrum in s1, the space s1, and the time instant at 30 minutes before. An arbitrary data handler can be written with these parameters.

get_spaces_every() is called on a spacetime to obtain a discrete set of spaces that meet a given spatiotemporal resolution. In Line 11, this method returns spaces at every five minutes with the 10 second time band, and each space contains data available on at least 80% of nodes within the space. Then, the maximum data is collected from each space (Lines 12 and 13). In Chronus, a list has the collect() method,[8] which takes a closure as its parameter, and the closure is executed on each element in a list. In this example, each element in spaces is passed to the space parameter of a closure (Line 14).

select() is used to obtain a subset of a list based on a certain condition specified in a closure. From Line 19 to 21, event_spaces obtains spaces, each of which yields 10 or lower standard deviation of data and finds an increase of 20 or more degrees in average in recent give minutes.

Event Detection with Chronus

An event detection application in Chronus pairs an *event specification* and a corresponding *event handler* to respond to the event. Listing 8.2 shows an example macroprogram that specifies an event detection application. It is visualized in Figure 8.2. Once an oil spill is detected, this macroprogram collects sensor data from an event area in the last 30 minutes and examines the source of the oil spill. The macroprogram also collects sensor data from the event area over the next one hour in a high spatiotemporal resolution to monitor the oil spill.

Listing 8.2: An Example Macro-Program for an Event-based Data Query

```
1   sp = Spacetime.new( GLOBALSPACE, Period.new( NOW, INF ) )
2   spaces = sp.get_spaces_every( Min 10, Sec 30, 100 )
3
4   event spaces {
5     sequence {
6       not get_data('f-spectrum', MAX) > 290);
7       get_data('f-spectrum', MAX) > 290;
8       get_data('droplet-concentration', MAX) > 10;
9       within MIN 30;
10    }
11    any {
12      get_data('f-spectrum', MAX) > 320;
13      window(get_data('d-concentration', MAX), AVG, HOUR -1) >
           15;
14    }
15    all {
16      get_data('f-spectrum', AVG) > 300;
17      get_data('d-concentration', AVG) > 20;
18    }
19  }
20  execute{ |event_space, event_time|
21    # query for the past
```

---

[8]In Ruby, a method can be called without parentheses when it takes no parameters.

```
22      sp1 = Spacetime.new(event_space, event_time, Min -30)
23      past_spaces = sp1.get_spaces_every(Min 6, Sec 20, 50)
24      num_of_nodes = past_spaces.get_nodes.select{ |node|
25          # @CWS_ROUTING
26          node.get_data('f-spectrum', Min 3) > 280}.size
27
28      # query for the future
29      s2 = Circle.new( event_area.centroid, event_area.radius *
            2 )
30      sp2 = Spacetime.new( s2, event_time, Hr 1 )
31      future_spaces = sp2.get_spaces_every( Min 3, Sec 10, 80 )
32      future_spaces.get_data( 'f-spectrum', MAX, Min 1 ){
33          | data_type, value, space, time |
34          # data handler }
35   }
36 }
```

An event detection is performed against a set of spaces, which is obtained by get_spaces_every(). Line 1 obtains a spacetime sp that starts from the current time and covers whole observation area. GLOBALSPACE is a special type of space that represents the whole observation area, and INF represents the infinite future. spaces in Line 2 represents a set of GLOBALSPACEs at every 10 minutes with the 30 second time band in the future.

An event specification (Lines 4 to 19) defines an event(s) (Lines 5 to 18) and an event handler (Line 20 to 35). Listing 8.3 shows the language syntax to define event specifications. An event specification is declared with the keyword event followed by a set of spaces (Line 4). Events are defined as the conditions to execute a corresponding event handler. Chronus supports three event types: *sequence*, *any*, and *all* events.

A *sequence* event fires when a set of *atomic events* occurs in a chronological order. An atomic event is defined with get_data() or window() operation. get_data() returns spatially processed data; that is, data aggregated over a certain space with one of the operators in Table 8.2. In Listing 8.2, a sequence event

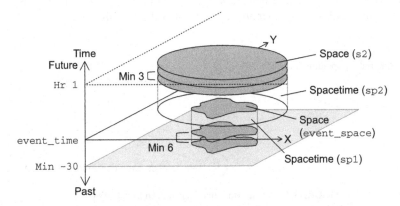

FIGURE 8.2 An Example Event-Based Data Query.

is defined with three atomic events (Lines 6 to 8). The sequence event fires if those atomic events occur chronologically in spaces. The first atomic event fires when the maximum fluorescence spectrum (f-spectrum) data does not exceed 290 (nm) (Line 6). The second atomic event fires when the maximum fluorescence spectrum (f-spectrum) data exceeds 290 (nm) (Line 7). The third one fires when the maximum concentration of oil droplet (d-concentration) exceeds 10 ($\mu L/L$) (Line 8). A sequential event can optionally specify a time constraint within which all atomic events occur. In Listing 8.2, three atomic events are specified to occur in 30 minutes (Line 9).

Listing 8.3: Event Specification Syntax

```
1  <query> = <event spec> <event handler>
2  <event spec> = 'event' <space> '{' {condition}+'}'
3  <condition> = <sequence> | <any> | <all>
4  <sequence> = 'sequence' '{' {<atomic event>}+ [<time
   restriction>] '}'
5  <atomic event> = (<get data> | <window>) <comparison operator
   > <number>
6  <time restriction> = 'within' <time>
7  <any> = 'any' '{' {<atomic event>}+ '}'
8  <all> = 'all' '{' {<atomic event>}+ '}'
9  <event handler> = 'execute' '{' <Chronus code> '}'
```

A sequence condition is transformed into a state machine in which each state transition corresponds to an atomic event. For example, a sequence in Listing 8.2 is transformed to a state machine in Figure 8.3. A is the initial state, and it transits to B when the first atomic event, not get_data("f-spectrum," MAX) > 290, occurs. The sequence event's handler is executed in D.

An *any* event fires when one of defined atomic events occurs. Listing 8.2 uses window() todefine an atomic event. window() returns temporally processed data; that is, data aggregated over time with one of the operators in Table 8.2. In Listing 8.2, an any event is defined with two atomic events (Lines 12 and 13). It fires and executes a corresponding event handler if one of the two atomic events

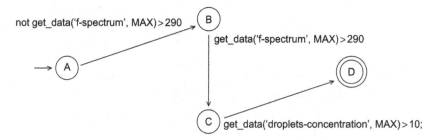

FIGURE 8.3 A State Transition Transformed from Listing 8.2.

occurs. The first atomic event fires when the maximum fluorescence spectrum (f-spectrum) data exceeds 320 (nm). The second atomic event fires when the average of the maximum oil droplet concentration (d-concentration) exceeds 15 ($\mu L/L$) in recent one hour.

An *all* event fires and executes a corresponding event hander when all of defined atomic events occur. It does not consider a chronological order of atomic events as a sequence event does. Listing 8.2 defines two atomic events (Lines 16 and 17).

An event handler is specified as a closure of execute (Lines 20 to 36). Its parameters are a space where an event has occurred and the time when the event has occurred (Line 20). In Line 23, a spacetime (sp1) is created to cover event_space over the past 30 minutes. Then, sp1 is used to examine how many nodes have observed 280 (nm) or higher fluorescence spectrum every six minutes. Line 25 is a special comment that contains an *attribute*, which starts with @. It is used to customize the default mapping from a macroprogram to a microprogram. The attribute @CWS_ROUTING declares to use an algorithm called CWS for packet routing. (See the later section Microprogramming QAs for more details about CWS.) In Line 36, another spacetime (sp2) is created to specify a data collection in the future. sp2 covers a larger space than event_space for an hour in the future. It is used to examine the maximum fluorescence spectrum every three minutes.

Listing 8.2 uses get_spaces_every() to obtain a set of spaces (Line 2). This operation guarantees that a Chronus application can watch a certain space with a certain spatiotemporal resolution. However, it does not suit for detecting highly critical events since an event may happen between two spaces in a list that get_spaces_every() returns and an event handler fails to respond to the event immediately. For example, when get_spaces_every() returns a set of spaces every 30 minutes, an application may respond to an event 30 minutes after the event happens. Reducing the interval of time to watch (i.e., the first parameter of get_spaces_every()), however, results in consuming unnecessarily large amount of battery power since get_spaces_every() may send packets to sensor nodes in a WSN to collect data (see details in Chronus Runtime Environment later in this chapter). In order to avoid this situation, Chronus supports get_spaces_upon() operation to obtain a set of spaces in addition to get_spaces_every(). Chronus assumes that each sensor node sends back data to a base station when a condition met (e.g., fluorescence spectrum exceeds 290(nm)), and get_spaces_upon() creates a space upon the arrival of data from sensor nodes.

Listing 8.4 shows that an example using get_spaces_upon() and get_spaces_every(). spaces1 contains a set of spaces created upon the arrival of data from nodes (Line 2), and spaces2 contains a set of spaces that satisfy a certain spatiotemporal resolution (Line 5). Chronus treats both types of spaces in the same way and allows for combining them (Line 7). With the combined spaces, observing_space, an application can monitor an area with a certain spatiotemporal resolution and respond to events that occur in the area immediately.

Listing 8.4: An Example Program Using get_spaces_upon()

```
1  sp = Spacetime.new( GLOBALSPACE , Period.new( NOW, INF ) )
2  spaces1 = sp.get_spaces_upon( Sec 30 )
3
4  sp = Spacetime.new( space , Period.new( NOW, INF ) )
5  spaces2 = sp.get_spaces_every( Min 10, Sec 30, 100 )
6
7  observing_space = spaces1 + spaces2
```

User-defined Data Aggregation Operators

Chronus allows application developers to introduce their own (i.e., user-defined) operators in addition to the predefined operators shown in Table 8.2. In Chronus, both predefined and user-defined operators are implemented in the same way.

Listing 8.5 shows the implementations of SUM, COUNT and AVG operators (Table 8.2). Each operator is defined as a *process object*, which is a code block that can be executed with the call() method. (See Line 14 for an example.) The keyword proc declares a process object, and its implementation is enclosed between the keywords do and end. sensor_readings is an input parameter (i.e., a set of sensor data to process) to each operator (Lines 1, 9, and 13).

Listing 8.6 shows an example user-defined operator, CENTROID, which returns the centroid of sensor data. In this way, developers can define and use arbitrary operators that they need in their applications

Listing 8.5: Predefined Data Aggregation Operators

```
1  SUM = proc do |sensor_readings|
2    sum = 0.0
3    sensor_readings.each do |sensor_reading|
4      sum += sensor_reading.value
5    end
6    sum
7  end
8
9  COUNT = proc do |sensor_readings|
10   sensor_readings.size
11 end
12
13 AVG = proc do |sensor_readings|
14   SUM.call(sensor_readings)/COUNT.call(sensor_readings)
15 end
```

Listing 8.6: An Example User-defined Operator for Data Aggregation

```
1  CENTROID = proc do |sensor_readings|
2    centroid = [0, 0] # indicates a coordinate (x, y)
3    sensor_readings.each do |sensor_reading|
4      centroid[0] += sensor_reading.value*sensor_reading.x
```

```
5 │ centroid[1] += sensor_reading.value*sensor_reading.y
6 │ end
7 │ centroid.map{ |value| value / sensor_readings.size }
8 │ end
```

CHRONUS IMPLEMENTATION

Chronus is currently implemented with an application architecture that leverages mobile agents in a push and pull hybrid manner (Figure 8.4). In this
architecture, each WSN application is designed as a collection of mobile agents,
and there are two types of agents: *event agents* and *query agents*. An event agent
(EA) is deployed on each node. It reads a sensor at every duty cycle and stores
its sensor data in a data storage on the local node. When an EA detects an event
(i.e., a significant change in its sensor data), it replicates itself, and a replicated
agent carries (or pushes) sensor data to a base station by moving in the network
on a hop-by-hop basis. Query agents (QAs) are deployed at Agent Repository
(Figure 8.4) and move to a certain spatial region (a certain set of nodes) to collect (or pull) sensor data that meet a certain temporal range. When EAs and
QAs arrive at the Chronus server, it extracts the sensor data the agents carry and
stores the data to a spatiotemporal database (STDB).

At the beginning of a WSN operation, the Chronus server examines network
topology and measures the latency of each link by propagating a measurement
message (similar to a hello message). EAs and QAs collect topology and latency
information as moving to base stations. When they arrive at the Chronus server,
they update the topology and latency information that the Chronus server maintains. The Chronus server also maintains each node's physical location through
a certain localization mechanism.

Visual Macroprogramming

In addition to textual macroprogramming shown in Figures 8.1 and 8.2, Chronus
provides a visual macroprogramming environment. It leverages Google Maps

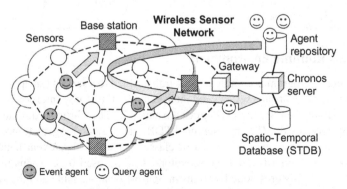

FIGURE 8.4 A Sample WSN Organization.

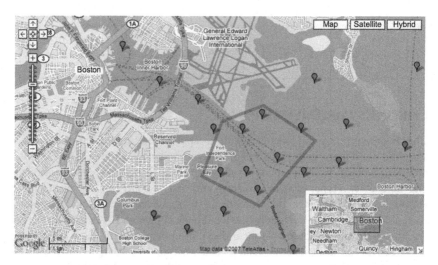

FIGURE 8.5 Chronus Visual Macroprogramming Environment.

(maps.google.com) to show the locations of sensor nodes as icons, and it allows application developers to graphically specify a space where they observe. Figure 8.5 shows a pentagonal space (an observation area) on an example WSN deployed at the Boston Harbor. Given a graphical space definition, the Chronus visual macroprogramming environment generates a skeleton macroprogram that describes a set of points (pairs of longitude and latitude) constructing the space. Listing 8.7 shows a macroprogram generated from a graphical space definition in Figure 8.5.

Listing 8.7: A Generated Skeleton Code

```
1  points = [ # ( Latitude , Longitude )
2    Point.new( 42.35042512243457 , -70.99880218505860 ),
3    Point.new( 42.34661907621049 , -71.01253509521484 ),
4    Point.new( 42.33342299848599 , -71.01905822753906 ),
5    Point.new( 42.32631627110434 , -70.99983215332031 ),
6    Point.new( 42.34205151655285 , -70.98129272460938 ) ]
7  s = Polygon.new( points )
```

Chronus Runtime Environment

Once a macroprogram is completed, it is transformed to a servlet (an application runnable on the Chronus server) and interpreted by the JRuby interpreter in the Chronus runtime environment (Figures 8.6 and 8.7). The Chronus runtime environment operates the Chronus server, STDB, gateway and Agent Repository. The Chronus library is a collection of classes, closures (data/event handlers) and process objects (user-defined operators) that are used by Chronus macroprograms. STDB stores node locations in SensorLocations table and the sensor

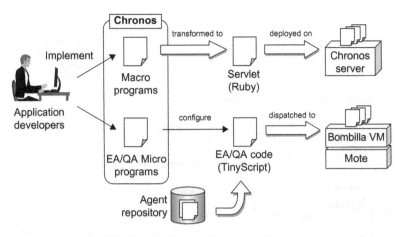

FIGURE 8.6   Chronus Development Process.

data agents carry in SensorData table. Node locations are represented in the OpenGIS Well-Known Text (WKT) format.[9]

When a Chronus macroprogram specifies a data query for the past, a SQL query is generated to obtain data from STDB. get_data() implements this mapping from a data query to SQL query. Listing 8.8 shows an example SQL query. It queries ids, locations, and sensor data from the nodes located in a certain space (space in Line 6). Contains() is an OpenGIS standard geographic function that examines if a geometry object (e.g., point, line and two dimensional surface) contains another geometry object. Also, this example query collects data from a given temporal domain (Lines 7 and 8). The result of this query is transformed to a Ruby object and is passed to a corresponding data handler in a macroprogram.

If STDB does not have enough data that satisfy a data query's spatiotemporal resolution, QAs are dispatched to certain sensor nodes in order to collect extra sensor data. They carry the data back to STDB.

When a Chronus macroprogram specifies a future data query, QAs are dispatched to a set of nodes that meet the query's spatial resolution. get_data() implements this mapping from a data query to QA dispatch. After a QA is dispatched to a node, the QA periodically collects sensor data in a given temporal resolution. It replicates itself when it collects data, and the replicated QA carries the data to STDB. The data is passed to a corresponding data handler.

As shown above, the notion of spacetime allows application developers to seamlessly specify data collection for the past and future. Also, developers do not have to know whether STDB has enough data that satisfy the spatiotemporal resolutions that they specify.

---

[9]www.opengeospatial.org

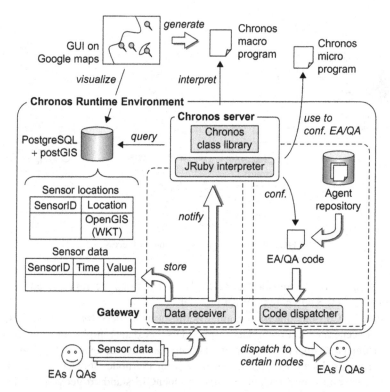

FIGURE 8.7 Chronus Runtime Environment.

Listing 8.8: An Example SQL

```
1  SELECT SensorLocations.id, SensorLocations.location ,
2         SensorData.value
3   FROM SensorLocations , SensorData
4  WHERE SensorLodations.id = SensorData.id AND
5         Contains(
6            space , SensorLocations.location ) = true AND
7         SensorData.time >= time - timeband AND
8         SensorData.time <= time + timeband;
```

In-Network Processing

As described earlier, Chronus macroprogramming language supports user-defined data processing operators. get_data() can specify a data processing operator as its parameter. Data processing is performed on the Chronus server or in a network depending on data queries.

When a data query collects sensor data in the past and STDB can provide enough data, collected data are processed on the Chronus server. Otherwise, a QA visits sensor nodes, collects sensor data, processes them on the last node of its route, and returns the result to a Chronus macro-program. This in-network data processing saves the power consumption in a sensor network by reducing the amount of data to exchange between nodes. In Chronus, to reduce the amount of data QAs brings, a QA is designed to have only its state and not to have code to execute on nodes. A code for in-network data processing is deployed only on the last node of QA's route, and a QA executes the code to process its data before returning to a base station. The Chronus server transforms a code for in-network data processing in Chronus macroprogramming language into TinyScript, and sends it to the last node of QA's route through the shortest path from a base station to the node before dispatching a QA.

Concurrency in the Chronus Server

Chronus macroprogramming language allows a Chronus macroprogram to have multiple data queries and data processing. This design strategy makes it easy to write queries and data processing, which depend on the results of preceding data queries and data processing. However, without an appropriate threading model (i.e., if Chronus macroprograms follow single thread model), they suffer from their low performance because data queries may take a long time and block other data queries and data processing continually. To maximize the performance of Chronus macroprograms, Chronus macroprograms automatically create new threads so that multiple data queries and data processing perform in a parallel manner.

Chronus macroprograms that deployed on the Chronus server, that is, servlets, can be invoked via SOAP, that is, an XML-based protocol [19]. As illustrated in Figure 8.8, a Chronus macro-program (Servlet) starts when its run() method is called. (run() method is automatically generated during a transformation from a Chronus macroprogram to a servlet, and it is used to execute the original Chronus macroprogram.) Then, a new thread (Data Collection Thread) is created when a Chronus macroprogram calls get_data() so that it can perform a data collection in parallel with the program's main thread. Each get_data() creates its own thread automatically. A data collection thread checks if STDB provides enough data and collects data from STDB or dispatches QAs. When a data collection thread dispatches QAs, it registers a corresponding event handler to a Chronus macroprogram. Once a gateway receives returning QA(s), it retrieves collected sensor data from the QA(s) and sends it to the Chronus server via SOAP (Figure 8.7). The Chronus server notifies it to a Chronus macroprogram, and a Chronus macroprogram invokes the registered event handler.

Since a program's main thread and data collection threads run in a parallel manner, get_data() may not be able to return a result to a program's main thread immediately. For example, in Listing 8.1, a variable max_values may not contain results of get_data() (Line 14) when a main thread accesses it (e.g., for drawing

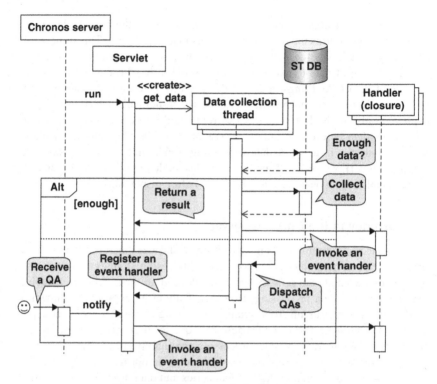

FIGURE 8.8   Concurrency in the Chronus Server.

a graph or creating another data query based on the variable). In Chronus, a main thread and a data collection thread are synchronized when a variable that contains a result of get_data() is accessed by a main thread.

CHRONUS MICROPROGRAMMING LANGUAGE

Chronus provides a *microprogramming* language to customize the default behavior of microprograms. While Chronus macroprograms are always written on spacetime, Chronus microprograms are always written on nodes. This allows developers to flexibly tune the performance and resource consumption of their applications by providing a means to tailor EAs and QAs. The Chronus microprogramming and macroprogramming languages share the same syntax and semantics.

Microprogramming EAs

By implementing a process object, the Chronus microprogramming language allows specifying a condition when an EA replicates itself and a replicated agent starts migrating to a base station. Listing 8.9 shows an example microprogram for EAs. Each node periodically obtains its sensor data and executes LOCAL_FILTERING process object. A process object returns true or false. When it returns true, an EA replicates itself on the node and a replicated EA starts migrating to a base station with sensor data. If it returns false, the node stores sensor data in its local storage (e.g., flush memory). A QA may visit the node and collect the stored data in future. In Listing 8.9, each node periodically check whether local sensor data exceed 300(nm) or not.

Listing 8.9: A Micro-Program for EAs (Local Filtering)

```
1  LOCAL_FILTERING = proc do |node|
2    node.get_data( 'f-spectrum' ) > 300
3  end
```

By leveraging attribute-oriented programming, Chronus allows use of microprograms in macroprograms while continuing to separate them. Listing 8.10 is an example Chronus macroprogram that refers a microprogram for EAs through the use of an attribute. A keyword starts with @ in a comment; for example, @LOCAL_FILTERING in Line 1, is called an attribute, and the attribute marks the following program element. In Listing 8.10, each node contained in the spacetime (sp) uses LOCAL_FILTERING (Listing 8.9) to decide when to send back data to a base station. If no attribute is specified right before an initialization of a spacetime, nodes in the spacetime do not send back data to a base station (only QAs collect data from nodes). Attribute-oriented programming improves the readability and maintainability of Chronus programs by separating macroprograms and microprograms clearly while providing a mechanism to combine them.

Listing 8.10: A Macroprogram for Deploying a Microprogram for EAs

```
1  # @LOCAL_FILTERING
2  sp = Spacetime.new( GLOBALSPACE , Period.new( Hr -1, Min 30))
3  spaces = sp.get_spaces_every( Min 10, Sec 30, 100 )
```

Listing 8.11 is another example of a microprogram for EAs. In Listing 8.11, when a local sensor data exceeds 300 (nm), each node obtains a list of neighbors within one hop away (Line 4) and checks if the average of their sensor data exceeds 300(nm). This algorithm reduces the number of false-positive sensor data compared with the algorithm in Listing 8.9 since each EA uses an average of neighbors' sensor data to decide whether to return a replicated EA, but may consume much energy since nodes exchange packets to obtain neighbors' sensor data.

Listing 8.11: A Micro-Program for EAs (Neighborhood Filtering)

```
1  NEIGHBORHOOD_FILTERING = proc do |node|
2    if node.get_data( 'f-spectrum' ) <= 300 then false
3
```

```
4    neighbors = node.get_neighbors_within(1)
5    total = node.get_data( 'f-spectrum' )
6    neighbors.each{ |neighbor|
7      total += neighbor.get_data( 'f-spectrum' ) }
8    total > 300 * (neighbors.size + 1);
9    end
```

Listing 8.12 is another example. In Listing 8.12, if local sensor data exceed 300(nm), each node broadcasts its local sensor data to one hop neighbors with a label f-spectrum (Lines 4 and 5). Once receiving a broadcast message, each node keeps it as a tuple consisting of a source node id and a received value in a table of which name is f-spectrum. After that, a node retrieves sensor data from its f-spectrum table and checks if the average of them exceeds 300 (nm). Compared with the algorithm in Listing 8.11, this algorithm consumes less energy since it uses broadcasts to exchange sensor data instead of node-to-node communications.

Listing 8.12: A Micro-Program for EAs (Gossip Filtering)

```
1    GOSSIP_FILTERING = proc do |node|
2      if node.get_data( 'f-spectrum' ) <= 300 then false
3
4      node.broadcast(
5        node.get_data( 'f-spectrum' ), 'f-spectrum', 1 )
6      total = node.get_data( 'f-spectrum' )
7      node.get_table( 'f-spectrum' ).each{ |node_id, value|
8        total += value }
9      total > 300 * (node.get_table.size + 1)
10   end
```

As shown in Listings 8.9, 8.11, and 8.12, Chronus microprogramming language allows defining arbitrary algorithm to decide when a replicated EA starts migrating to a base station. Depending on the requirements of WSN applications, for example, low latency, low energy consumption, or less false-positive sensor data, application developers can implement their own algorithms and deploy them on nodes in a certain space.

Implementation of EAs

Chronus extends a Bombilla VM [20] and TinyScript to support mobile agents as one of the messages that can move among sensor nodes with sensor data. A microprogram is used to configure EA code (template) in Agent Repository, and a configured EA code is deployed on certain nodes by the Chronus server (Figures 8.6 and 8.7). Listing 8.13 is an example EA code configured with Listing 8.9. A microprogram for EAs is transformed into TinyScript and copied to a template. The Chronus server deploys this code on certain nodes in a space specified in a Chronus macroprogram.

Listing 8.13: A Fragment of EA Code in TinyScript

```
1  agent ea;
2  private data = get_sensor_data();
3  if (get_sensor_data() > 300) then
4    ea = create_event_agent();
5    set_source(ea, id());
6    set_sensor_data(ea, data);
7    return_to_basestation(ea);
8  end if
```

Microprogramming QAs

In a default algorithm for QA's routing, only one QA visits to nodes in a certain space in the order of node's id. However, Chronus microprogramming language allows implementing QA's routing algorithms such as Clarke-Wright Savings (CWS) algorithm [21].

CWS is a well-known algorithm for Vehicle Routing Problem (VRP), one of NP-hard problems. The CWS algorithm is a heuristic algorithm that uses constructive methods to gradually create a feasible solution with modest computing cost. Basically, the CWS algorithm starts by assigning one agent per vertex (node) in the graph (sensor network). The algorithm then tries to combine two routes so that an agent will serve two vertices. The algorithm calculates the "savings" of every pair of routes, where the savings is the reduced total link cost of an agent after a pair of route is combined. The pair of routes that have the highest saving will then be combined if no constraint (e.g., deadline) is violated.

Listing 8.14 implements a QA's routing algorithm based on CWS. CWS_ROUTING is a process object that is executed right before dispatching QAs by the Chronus server. The process object takes a set of nodes to visit (nodes in Line 2), a spatial resolution and a tolerable delay specified by a data query (percentage and tolerable_delay), and the maximum number of nodes an agent can visit (max_nodes). Since the size of the agent's payload is predefined, an agent is not allowed to visit and collect data from more than a certain number of nodes. The process object returns a set of sequences of nodes as routes on which each QA follows (routes in Line 9); for example, if it returns three sequences of nodes, three QAs will be dispatched, and each of them uses each sequence as its route. Moreover, a process object returns a set of sequences of nodes, a QA replicates itself on an intermediate node and visit nodes in parallel. For example, when a process returns a set of two sequences of nodes as {{5, 9, 10}, {5, 7}}, a QA moves from a base station to node 5 and replicates itself. One QA visits to nodes 9 and 10, and the other visits to node 7. After that, the two QAs merge into one QA, and it returns to a base station.

In Listing 8.14, CWS_ROUTING selects part of nodes based on a spatial resolution (Lines 9 to 12) and calculates the savings of each adjacent nodes pair (Line 14 to Line 22). After that, routes are created by connecting two adjacent nodes in the order of savings. As described earlier, the Chronus server stores the topology and latency information collected by EAs and QAs, and microprograms can use that information through node object, for example, get_closest_node(), get_shortest_path() and get_delay() methods (Line 4 to 6).

Listing 8.14: A Micro-Program for QAs (CWS Routing)

```
1   CWS_ROUTING = proc do
2   | nodes, percentage, tolerable_delay, max_nodes |
3
4   closest = get_closest_node( base, nodes )
5   delay = tolerable_delay/2 -
6    closest.get_shortest_path(base).get_delay(base)
7
8   # select closest nodes
9   nodes = nodes.sort{|a, b|
10   a.get_shortest_path(closest).get_delay <=>
11   b.get_shortest_path(closest).get_delay}
12   [0, (nodes.length * percentage/100).round - 1]
13
14   nodes.each{ |node1|   # get savings of each pair
15   nodes.each{ |node2|
16    next if node1.get_hops(node2) != 1
17    saving =
18     node1.get_shortest_path(closest).get_delay +
19     node2.get_shortest_path(closest).get_delay -
20     node1.get_shortest_path(node2).get_delay
21    savings[saving].push({node1, node2}) } }
22
23   # connect nodes in the order of savings
24   savings.keys.sort{ |saving|
25    savings[saving].each{ |pair|
26    if !pair[0].in_route && !pair[1].in_route ||
27    pair[0].is_end != pair[1].is_end then
28     route1 = pair[0].get_route_from(closest)
29     route2 = pair[1].get_route_from(closest)
30     if route1.get_delay <= delay &&
31     route1.get_size <= max_nodes &&
32     route2.get_delay <= delay &&
33     route2.get_size <= max_nodes then
34     pair[0].connect_with(pair[1]) # connect
35    end
36   end } }
37
38   # return routes
39   nodes.select{|node| node.is_end}
40    .map{|node| node.get_route_from(closest)}
41   end
```

In addition to microprograms for EA as described earlier, microprograms for QA can be referred to in macroprograms through the use of attributes.

Listing 8.15 is an example Chronus macroprogram that refers a microprogram for QAs through the use of an attribute. A microprogram for QAs is used as a default in a spacetime when a corresponding attribute marks an initialization of the spacetime (Lines 1 and 2). Also, a microprogram for QAs is used only for performing a certain get_data() when a corresponding attribute marks a get_data() method call (Line 1 and 2).

Listing 8.15: A Macroprogram for Deploying a Microprogram for QAs

```
1   # @CWS_ROUTING
2   sp = Spacetime.new( GLOBALSPACE, Period.new( Hr -1, Min 30 ))
3   spaces = sp.get_spaces_every( Min 10, Sec 30, 100 )
4
5   max_values = spaces.collect { |space|
6     # @CWS_ROUTING
7     space.get_data( 'f-spectrum', MAX, Min 2 ){
8       | data_type, value, space, time |
9       # data handler
10    }
11  }
```

Implementation of QAs

A microprogram for QAs is executed on the Chronus server to configure QAs' routes. Each QA is implemented in TinyScript. As illustrated in Figures 8.6 and 8.7, QA's template code is stored in Agent Repository. The Chronus server configures QA's route with a microprogram.

Listing 8.16 is a fragment of a configured QA code in TinyScript that is executed once at a base station. set_agent_path() sets a path, that is, a sequence of nodes to visit (Lines 5 and 6). set_start_collecting() sets when to start collecting data by specifying an index of a node (Line 7). If multiple QAs are used to collect data in parallel, each QA's route is specified in the sequence delimited with 0. In this example, a QA migrates from a base station to node 1 and 3, and starts collecting data. At node 3, a QA creates another QA, and one QA collects data from nodes 11 and 9, and another QA collects data from nodes 10 and 13 (Line 5). After visiting all nodes, each QAs returns to the node they split, node 3 in this example, and merge themselves. In this example, two QAs (the first and the second QAs) are merged into one QA (Lines 9 and 10). A list of nodes to collect data is provided by a microprogram for QAs. A list of nodes before starting a data collection (nodes 1 and 3) is the shortest path from a base station to the node to start collecting sensor data (node 12). Also, set_timestamp_from() and set_timestamp_untill() specifies a time window of data to collect. Chronus assumes timers of all nodes are synchronized and the Chronus server can convert a representation of a time instant in a macroprogram, that is, absolute and relative times, into a clock of node.

Listing 8.16: A Fragment of QA Code

```
1   agent qa;
2   buffer path;
3   buffer merge;
4   qa = create_query_agent();
5   path[]=1; path[]=3; path[]=11; path[]=9; path[]=0; path[]=10;
        path[]=13;
6   set_agent_path(qa, path);
7   set_start_collecting(qa, 1);
8   set_agent_num_of_qa(qa, 2);
9   merge[] = 1; merge[] = 2;
10  set_agents_merge(qa, merge);
11  set_timestamp_from(qa, 100);
12  set_tiemstamp_untill(qa, 500);
13  migrate(qa);
```

Listing 8.17 is a fragment of a code deployed on each node beforehand and is used to accept QAs. It is executed when a node receives a broadcast message. (QAs are transmitted via broadcast.) It checks whether a QA collects data from the current node (Line 6). If the current node is the last one to visit, a QA executes a code for in-network processing and returns to a base station along the shortest path (Lines 11 and 12). If not, a QA migrates to the next node (Line 14).

Listing 8.17: A Fragment of Code to Accept EAs

```
1   agent qa, copy;
2   buffer path;
3   private node_id;
4   qa = migratebuf();   # retrieves a QA from a buffer
5   node_id = id();       # get the current node id
6
7   # start collecting data?
8   if start_collection(qa, node_id) then
9     for i = 1 to get_num_of_qa(qa) - 1
10      copy = create_query_agent();
11      path = get_path(copy, i);
12      set_agent_path(copy, path);
13      set_start_collecting(copy, 1);
14      migrate(copy);
15    next i
16  end if
17
18  if (do_collection(qa, node_id)) then   # collect data?
19    add_data(qa, get_sensor_data());
20  end if
21
22  if (is_end(qa, node_id)) then # the last node to visit?
23    # do in-network processing here
```

```
24 |   return_to_basestation(qa);
25 | else
26 |   migrate(qa);   # move to the next node
27 | end if
```

SIMULATION EVALUATION

This simulation study simulates a WSN deployed on the sea to detect oil spills in the Boston Harbor of Massachusetts (Figure 8.9). The WSN consists of nodes equipped with fluorometers. Nodes are deployed in a 6×7 grid topology in an area of approximately 620×720 square meters. They use MICA2 motes with the outdoor transmission range (radius) of 150 meters, 38.4 kbps bandwidth, 128 kB program memory (flush memory) and 2000 mAh battery capacity (two AA battery cells). The node running one of four WSN corners works as the base station. This study assumes that 100 barrels (approximately 4,200 gallons) of crude oil is spilled at the center of WSN. A simulation data set is generated with an oil spill trajectory model implemented in the General NOAA Oil Modeling Environment [22]. Sensor data shows a fluorescence spectrum of 280 (nm) when there is no spilled oil, and it reaches 318 (nm) when there exists oil. Each sensor has a white noise that is simulated as a normal random variable with its mean of zero and standard deviation of 5% of sensor data.

Event Detection

This section describes the performance differences between EA's algorithms shown in Listings 8.9 (Local Filtering), 8.11 (Neighborhood Filtering), and 8.12 (Gossip Filtering). Figures 8.10 and 8.13 show the number of packets needed

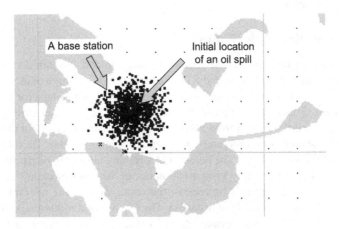

FIGURE 8.9 A Simulated Oil Spill.

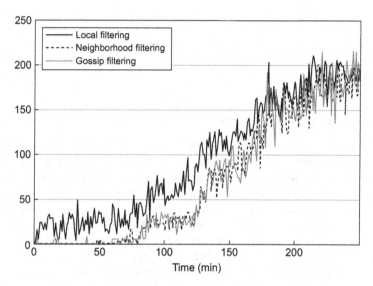

FIGURE 8.10   Packet Transmission.

to transmit EAs to the base station and the number of false-positive data. With Local Filtering, nodes decide whether to send replicated EAs independently; the base station receives many false-positive data. With Neighborhood Filtering and Gossip Filtering, the base station receives much less false-positive data because nodes interact with each other before sending EAs. However, as shown in Figure 8.11, this interaction requires control overhead (i.e., power consumption). (There is no control overhead in Local Filtering.) Figure 8.12 shows the total number of packet transmissions. By reducing the number of false-positive data, the total number of packet transmissions is comparable in Gossip Filtering and Local Filtering.

Data Collection in the Future

As described earlier, QAs are dispatched to nodes to collect senor data when a query retrieves historical data and STDB cannot provide enough data.

Table 8.3 compares the behavior of different routing algorithms for QA, that is, a default QA's routing algorithm and the CWS algorithm in Listing 8.14, when a query retrieves data from nodes in 3×3 nodes in the center of the WSN with 100% spatial resolution and three minutes tolerable delay. Each QA can contain 13 sensor readings. The default QA's routing algorithm dispatches only one QA, and the QA simply visits to all nodes in the order of node's id. Since it does not consider query's timeliness, the result violates the tolerable delay specified by a query (i.e., three minutes). The CWS-based routing algorithm in the earlier section Microprogramming QAs considers the tolerable delay and dispatches three QAs simultaneously, and one of QAs takes 2,887 ms to collect

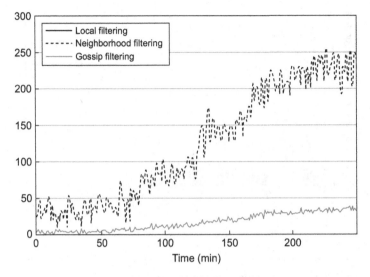

FIGURE 8.11   Control Overhead.

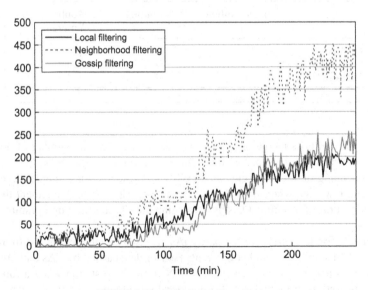

FIGURE 8.12   Total Packet Transmission.

data and return to the base station. (The other two take 2,679 ms and 2,380 ms). Since the routes of these three QAs are partially overwrapped, especially a route between a base station and the area where the 3×3 nodes are located, the total number of hops QAs take (battery consumption) is larger than one of the default routing algorithm. When an extended version of CWS-based routing algorithm is used, the total number of hops drops significantly since a QA replicates itself

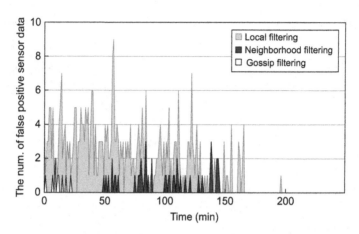

FIGURE 8.13   The Number of False Positive Data.

TABLE 8.3 A Measurement on QA for the Past

	Default Routing	CWS Routing	CWS Routing with QA merging
# of QAs	1	3	1 (split into 3)
Latency (ms)	4459	2887	2887
Total # of Hops	26	48	28

and replicated QAs merge on an intermediate node. The total number of hops QAs take is almost the same as one of the default routing algorithms.

Depending on the requirements of WSN applications, for example, timeliness and low energy consumption, application developers can implement their own algorithms for routing QAs by leveraging the Chronus microprogramming language.

In addition to queries for the past, QAs are used for queries for the future. Figure 8.14 shows the number of agents (sensor data carried by EAs and QAs) from 3×3 nodes in the center of the WSN to the base station when a future query is used as in Listing 8.2. The temporal resolution of the future query is three minutes, that is, obtain data every three minutes, and the spatial resolution varies from 0 to 100%. In addition to QAs, Gossip Filtering-based EAs are deployed on each node.

When a spatial resolution is 0%, no future query is used, and few sensor data are transmitted during the first 75 minutes since only EAs send sensor data (replicated EAs). When a spatial resolution is larger than 0%, deployed QAs send sensor data (replicated QAs) to the base station every three minutes according to a spatial resolution even if deployed EAs do not send sensor data.

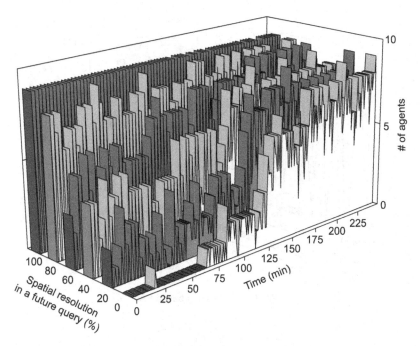

FIGURE 8.14   The Number of Sensor Data Received by the Base Station with a Future Query.

In Figure 8.14, spikes appear every three minutes, and they corresponds to transmitted QAs. In this way, a future query in Chronus allows collecting sensor data to satisfy specified spatiotemporal resolutions even if there is no event.

Data Collection in the Past

This section describes differences in the response time of data queries that retrieve data in the past from nodes in 3×3 nodes in the center of the WSN with 100% spatial resolution (Figure 8.15(a) and 8.15(b)). An application executes data queries, get_data() method call with three minutes tolerable delay, 240 minutes after an oil spill occurs. -30 min means that an application examines the data at 30 minutes before (at 210 minutes after an oil spill occurs). If every node in the 3×3 area sent EAs to a base station, the response time of data queries would be almost zero because STDP can provide enough data to an application. However, if STDP cannot provide enough data, Chronus runtime automatically dispatches QAs to sensor nodes to collect data. Since QAs use the CWS algorithm defined in Microprogramming QAs and data query's tolerable delay is three minutes, it takes 2,887 ms when no data is available in STDB. Also, it takes 1,291 ms for a QA to visit only the closest node in the 3×3 area.

Figures 8.15(a) and 8.15(b) show response times when EAs work with Gossip Filtering and Local Filtering algorithms, respectively. Response times

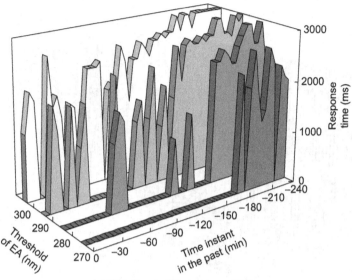

(a) With Gossip Filtering-based EAs

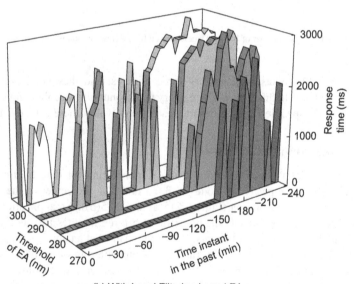

(b) With Local Filtering-based EAs

FIGURE 8.15 Response Time of Data Queries.

in Figure 8.15(b) are shorter than those in Figure 8.15(a) since Local Filtering-based EAs send much sensor data, including false-positive data, than Gossip Filtering-based EAs as shown in the earlier section, Event Detection. When an application gives weight to data collection rather than data detection, Local

Filtering-based EAs give a positive impact since it can reduce the response time of data queries for the past. When an application gives weight to data detection rather than data collection, Gossip Filtering-based EA has a positive impact since it reduces the number of false-positive data and consumes less battery power. As the results show, Chronus allows developers to flexibly tune their applications through the use of microprogramming languages.

Line of Code

This section compares lines of code (LOC) of applications in Chronus and nesC. A number of WSN applications are currently implemented in nesC [23], a dialect of the C language, and deployed on the TinyOS operating system [24], which provides low-level libraries for basic functionalities such as sensor reading and node-to-node communication. nesC and TinyOS hide hardware-level details; however, they do not help developers to rapidly implement their applications because applications in nesC are required to handle low-level mechanisms (e.g., memory management and routing).

Listing 8.18 is a Chronus code to collect data from sensor nodes by leveraging QAs. The code hides (1), which routes QAs use to move in WSNs and (2) how each node routes QAs. As described earlier, Chronus allows reuse of microprograms by leveraging an attribute-oriented technique. Listing 8.18 hides 33 lines of code defined in Listing 8.14. Moreover, application developers are required to implement nesC code deployed on each sensor node to route QAs. Listing 8.19 shows a fragment of the code. ReceiveMsg.receive() is invoked when a node receives a QA (Line 9). After that, a code collects sensor data in the past, appends the data to a QA, and routes the QA to the next node. In addition to a code in Listing 8.19, a code for configure modules (e.g., bind the QARouting module with a module that implements ReceiveMsg interface) is required to be implemented. The total LOC of nesC code is approximately 70.

Listing 8.18: Chronus Code to Collect Data Using QAs

```
1  # @CWS_ROUTING
2  sp = Spacetime.new( GLOBALSPACE , Period.new( Hr -1, Min 30))
3  spaces = sp.get_spaces_every( Min 10, Sec 30, 100 )
```

Listing 8.19: nesC version QA Routing Module

```
1  module QARouting {
2    uses {
3      interface ReceiveMsg;
4      interface SendMsg;
5      interface StdControl;
6  } }
7  implementation {
8    // invoked once a node receive a QA
```

```
 9    event TOS_MsgPtr ReceiveMsg.receive(TOS_MsgPtr m) {
10      AgentMsg* agent = (AgentMsg*)m->data;
11
12      // collect sensor data from the current node
13      if(agent ->isReturning == FALSE &&
14         agent ->startCollectionIndex >= agent ->numOfHops){
15        uint16_t data = getDataAt(agent ->data_timestamp);
16        append(agent , TOS_LOCAL_ADDRESS , data);
17      }
18
19      // check if the current node is the last one to visit
20      if( getLastNode(agent) == TOS_LOCAL_ADDRESS )
21        agent ->isReturning = TRUE;
22
23      // go to the next node
24      agent ->numOfHops++;
25      uint16_t nextNode = call getNextNode(agent);
26      if(call Send.send(nextNode , sizeof(AgentMsg), &m) !=
         SUCCESS){
27        return FAIL;
28      }
29      return SUCCESS;
30    }
31
32    // get sensor data in the past from a log
33    static uint16_t getDataAt(uint16_t timestamp){...}
34    // add a new data to an agent
35    static uint16_t append(AgentMsg* agent , uint16_t addr ,
         uint16_t data){...}
36    // get the last node to visit
37    static uint16_t getLastNode(AgentMsg* agent){...}
38    // get the next node to visit
39    static uint16_t getNextNode(AgentMsg* agent){...}
40    }
```

Chronus microprograms also hide the details of applications running on each sensor node as well as macroprograms do. Listing 8.20 shows a fragment of nesC version of Local-Filtering algorithm. ADC.dataReady() is invoked right after an Analog-Digital Converter module provides sensor data to an application. If the data exceeds a certain threshold (Line 5), it calls SendData() to send the data to a base station (Line 7). The LOC of Chronus version of Local-Filtering algorithm is only three (Listing 8.9), while one of nesC version is 20.

Listing 8.20: nesC version Local-Filtering Algorithm

```
1    // invoked once an AD converter is ready
2    async event result_t ADC.dataReady(uint16_t data){
3      atomic {
4        if (!gfSendBusy) {
5          gfSendBusy = TRUE;
6          if(data > SENSOR_READING_THRESHOLD) {
```

```
7         post SendData( data );
8       } } }
9       return SUCCESS;
10      }
11
12      task void SendData(uint16_t data){
13        EventAgentMsg *pReading;
14        if (pReading = (EventAgentMsg *)call Send.getBuffer(&
          gMsgBuffer ,&Len)) {
15          pReading ->type = F_SPECTRUM;
16          pReading ->parentaddr = call RouteControl.getParent();
17          pReading ->reading = data;
18          // send a message to a base station
19          if ((call Send.send(&gMsgBuffer ,sizeof(EventAgentMsg)))
            != SUCCESS){
20            atomic gfSendBusy = FALSE;
21          }
22      } }
```

As the results show, Chronus reduces the LOC of WSN applications significantly and allows developers to implement their applications rapidly.

Memory Footprint

Table 8.4 shows the memory footprint of microprograms deployed on a sensor node. The total footprint counts the memory consumption of TinyOS, Bombilla VM, EA code, and QA code. As shown in the table, Chronus's microprograms are lightweight enough to run on a MICA2 node, which has 128 KB memory space. It can operate on a smaller-scale sensor node; for example, TelosB, which has 48 KB memory space.

RELATED WORK

This work is a set of extensions to the authors' previous work [25, 26]. This chapter investigates several extensions to [25]; for example, user-defined operators and microprogramming. This chapter also extends [26] to study complex event specification in macroprograms, streamlined mapping from macroprograms to microprograms based on the notion of attribute-oriented

TABLE 8.4 Memory Footprint

EA Algorithms	Total Footprint (KB)	EA Code Footprint (KB)
Local Filtering	41.3	0.077
Neighborhood Filtering	41.3	0.114
Gossip Filtering	41.3	0.116

programming, and extra routing optimization for QAs. Moreover, this chapter provides extended simulation results to evaluate the performance and resource consumption of WSN applications built with Chronus.

Pleiades [27] and Snlog [28] are the languages for spatial macroprogramming. They provide programming abstractions to describe spatial relationships and data aggregation operations across nodes. Event detection can be expressed without specifying the low-level details of node-to-node communication and data aggregation. However, these languages require application developers to explicitly write programs to individual nodes. For example, they often need to access the states of individual nodes. In contrast, Chronus allows developers to program event detection to spacetime as a global behavior of each application. Also, Kairos and SNLong do not consider the temporal aspect of sensor data. They do not support data collection for the past; all sensor data are always considered as data collected at the current time frame.

Proto [29], Regiment [30], and Flask [31] are similar to Chronus in that they support in-network data processing and spatiotemporal event detection. While they allow developers to specify arbitrary event detection algorithms, they do not support data collection and the notion of spatial and temporal resolutions. Chronus supports data collection for the future and the past in arbitrary spatiotemporal resolutions.

TinyDB [32] and Semantic Stream [33] perform in-network data processing as well as spatiotemporal data collection by extending SQL and Prolog, respectively. They aid in program data collection for the future, but not for the past. Moreover, their expressiveness is too limited to implement data handlers, although they are well applicable to specify data queries. Therefore, developers need to learn and use extra languages to implement data handlers. In contrast, Chronus supports spatiotemporal data collection for the future and the past. Its expressiveness is high enough to provide integrated programming abstractions for data queries and data handlers. Also, TinyDB supports event-based data collection that is executed upon a predefined event; however, it does not support event detection on individual nodes.

SPIRE [34] and SwissQM [35] propose SQL-based languages for complex event detection in RFID systems and WSNs, respectively. GEM [36] proposes a Petri Net-based visual language to detect complex events. The expressiveness of these languages is too limited to implement event handlers, although they are well applicable to define event specifications. Chronus's expressiveness is higher to provide integrated programming abstractions for event specifications and event handlers.

This work is the first attempt to investigate a push-pull hybrid WSN architecture that performs spatiotemporal event detection and data collection. Most of existing push-pull hybrid WSNs do not address the spatiotemporal aspects of sensor data [37–39]. They also assume static network structures and topologies (e.g., star and grid topologies). Therefore, data collection can be fragile against node/link failures and node addition/redeployment. In contrast, Chronus can operate in arbitrary network structures and topologies. It can implement failure-resilient queries by having the Chronus server dynamically adjust the migration route that each QA follows.

PRESTO performs push-pull hybrid event detection and data collection in arbitrary network structures and topologies [40]. While it considers the temporal aspect in data queries, it does not consider their spatial aspect. Moreover, it does not support data collection in the future as well as in-network data processing.

CONCLUSION

This chapter proposes a new programming paradigm for autonomic WSNs, spatiotemporal macroprogramming. It is designed to reduce the complexity of programming spatiotemporal event detection and data collection. This chapter discusses Chronus's design, implementation, runtime environment, and performance implications.

Chronus is currently limited in two-dimensional physical space to deploy sensor nodes. Plans are to extend it for using three-dimensional physical space (i.e., four-dimensional spacetime). Accordingly, new applications will be studied, such as three-dimentional building monitoring and atmospheric monitoring.

REFERENCES

[1] I.F. Akyildiz, W. Su, Y. Sankarasubramaniam, and E. Cayirci, "A survey of sensor network applications," *IEEE Communications Magazine* Vol. 40, No. 8, pp. 102–114, 2002.

[2] T. Arampatzis, J. Lygeros, and S. Manesis, "A survey of applications of wireless sensors and wireless sensor networks," IEEE International Symposium on Intelligent Control, 2005.

[3] D. Estrin, R. Govindan, J. Heidemann, and S. Kumar, "Next century challenges: Scalable coordination in sensor networks," ACM International Conference on Mobile Computing and Networks, 1999.

[4] U.S. Coast Guard, Polluting Incident Compendium: Cumulative Data and Graphics for Oil Spills 1973–2004 (September 2006).

[5] L. Siegel, Navy oil spills, http://www.cpeo.org (November 1998).

[6] J.M. Andrews and S.H. Lieberman, "Multispectral fluorometric sensor for real time in-situ detection of marine petroleum spills," The Oil and Hydrocarbon Spills, Modeling, Analysis and Control Conference, 1998.

[7] C. Brown, M. Fingas, and J. An, "Laser fluorosensors: A survey of applications and developments," The Arctic and Marine Oil Spill Program Technical Seminar, 2001.

[8] R.L. Mowery, H.D. Ladouceur, and A. Purdy, Enhancement to the ET-35N Oil Content Monitor: A Model Based on Mie Scattering Theory, Naval Research Laboratory (1997).

[9] J.A. Nardella, T.A. Raw, and G.H. Stokes, "New technology in oil content monitors," *Naval Engineers Journal*, Vol. 101, No. 2, pp. 48–55, 1989.

[10] U.S. Environmental Protection Agency, "Understanding oil spills and oil spill response" (1999).

[11] M.F. Fingas and C.E. Brown, "Review of oil spill remote sensors," International Conference on Remote Sensing for Marine and Coastal Environments, 2002.

[12] G. Booch, *Object-Oriented Analysis and Design with Applications*, 2nd ed., Addison-Wesley, 1993.

[13]   E. Meijer and P. Drayton, "Static typing where possible, dynamic typing when needed: The end of the cold war between programming languages," ACM SIGPLAN Conference on Object-Oriented Programming, Systems, Languages, and Applications, Workshop on Revival of Dynamic Languages, 2004.

[14]   J.K. Ousterhout, "Scripting: Higher level programming for the 21st century," IEEE Computer, Vol. 31, No. 3, pp. 23–30, 1998.

[15]   M. Mernik, J. Heering, and A. Sloane, "When and how to develop domain-specific languages," ACM Comp. Surveys, Vol. 37, No. 4, pp. 316–344, 2005.

[16]   J.S. Cuadrado and J. Molina, "Building domain-specific languages for model-driven development," IEEE Software, Vol. 24, No. 5, pp. 48–55, 2007.

[17]   A. Bryant, A. Catton, K.D. Volder, and G. Murphy, "Explicit programming," International Conference on Aspect-Oriented Software Development, 2002.

[18]   H. Wada, J. Suzuki, K. Oba, S. Takada, and N. Doi, "mturnpike: A model-driven framework for domain specific software development," JSST Computer Software, Vol. 23, No. 3, pp. 158–169, 2006.

[19]   T.W.W.W. Consortium (Ed.), SOAP Version 1.2, 2003.

[20]   P. Levis and D. Culler, "Mate: A tiny virtual machine for sensor networks," ACM International Conference on Architectural Support for Programming Languages and Operating Systems, 2002.

[21]   G. Clarke and J.W. Wright, "Scheduling of vehicles from a central depot to a number of delivery points," Operations Research, Vol. 4, No. 12, pp. 568–581, 1964.

[22]   C. Beegle-Krause, "General noaa oil modeling environment (gnome): A new spill trajectory model," International Oil Spill Conference, 2001.

[23]   D. Gay, P. Levis, R. Behren, M. Welsh, E. Brewer, and D. Culler, "The nesc language: A holistic approach to networked embedded systems," ACM Conference on Programming Language Design and Implementation, 2003.

[24]   J. Hill, R. Szewczyk, A. Woo, S. Hollar, D.E. Culler, and K.S.J. Pister, "System architecture directions for networked sensors, " ACM International Conference on Architectural Support for Programming Languages and Operating Systems, 2000.

[25]   H. Wada, P. Boonma, and J. Suzuki, "Macroprogramming spatio-temporal event detection and data collection in wireless sensor networks: An implementation and evaluation study, " Hawaii International Conference on System Sciences, 2008.

[26]   H. Wada, P. Boonma, and J. Suzuki, "A spacetime oriented macro programming paradigm for push-pull hybrid sensor networking," IEEE ICCC, Workshop on Advanced Networking and Communications, 2007.

[27]   N. Kothari, R. Gummadi, T. Millstein, and R. Govindan, "Reliable and efficient programming abstractions for wireless sensor networks," SIGPLAN Conference on Programming Language Design and Implementation, 2007.

[28]   D. Chu, L. Popa, A. Tavakoli, J. Hellerstein, P. Levis, S. Shenker, and I. Stoica, "The design and implementation of a declarative sensor network system" ACM Conference on Embedded networked Sensor Systems, 2007.

[29]   J. Beal and J. Bachrach, "Infrastructure for engineered emergence on sensor/actuator networks," IEEE Intelligent Systems, Vol. 21, No. 2, pp. 10–19, 2006.

[30]   R. Newton, G. Morrisett, and M. Welsh, "The regiment macroprogramming system," International Conference on Information Processing in Sensor Networks, 2007.

[31]   G. Mainland, G. Morrisett, and M. Welsh, "Flask: Staged functional programming for sensor networks," ACM SIGPLAN International Conference on Functional Programming, 2008.

[32] S. Madden, M. Franklin, J. Hellerstein, and W. Hong, "Tinydb: An acqusitional query processing system for sensor networks," *ACM Transactions on Database Systems*, Vol. 30, No. 1, pp. 122–173, 2005.

[33] K. Whitehouse, J. Liu, and F. Zhao, "Semantic streams: A framework for composable inference over sensor data," European Workshop on Wireless Sensor Networks, 2006.

[34] R. Cocci, Y. Diao, and P. Shenoy, "Spire: Scalable processing of rfid event streams," RFID Academic Convocation, 2007.

[35] R. Müller, G. Alonso, and D. Kossmann, "Swissqm: Next generation data processing in sensor networks," Conference on Innovative Data Systems Research, 2007.

[36] B. Jiao, S. Son, and J. Stankovic, "Gem: Generic event service middleware for wireless sensor networks," International Workshop on Networked Sensing Systems, 2005.

[37] W. Liu, Y. Zhang, W. Lou, and Y. Fang, "Managing wireless sensor networks with supply chain strategy," International Conference on Quality of Service in Heterogeneous Wired/Wireless Networks, 2004.

[38] W. C. Lee, M. Wu, J. Xu, and X. Tang, "Monitoring top-k query in wireless sensor networks," IEEE International Conference on Data Engineering, 2006.

[39] S. Kapadia and B. Krishnamachari, "Comparative analysis of push-pull query strategies for wireless sensor networks," International Conference on Distributed Computing in Sensor Systems, 2006.

[40] M. Li, D. Ganesan, and P. Shenoy, "Presto: Feedback-driven data management in sensor networks," ACM/USENIX Symposium on Networked Systems Design and Impl., 2006.

Chapter 9

Security Metrics for Risk-aware Automated Policy Management

E. Al-Shaer, L. Khan, M. S. Ahmed, and M. Taibah

ABSTRACT

The factors on which security depends are of a dynamic nature. They include emergence of new vulnerabilities and threats, policy structure, and network traffic. Therefore, evaluating security from both the service and policy perspective can allow the management system to make decisions regarding how a system should be changed to enhance security as par the management objective. Such decision making includes choosing between alternative security architectures, designing security countermeasures, and systematically modifying security configurations to improve security. Moreover, this evaluation must be done dynamically to handle real-time changes to the network threat. In this chapter, we provide a security metric framework that quantifies objectively the most significant security risk factors, which include the historical trend of vulnerabilities of the remotely accessible services, prediction of potential vulnerabilities in the near future for these services and their estimated severity, existing vulnerabilities, unused address space, and finally propagation of an attack within the network. These factors cover both the service aspect and the network aspect of risk toward a system. We have implemented this framework as a user-friendly tool called *Risk-based prOactive seCurity cOnfiguration maNAger (ROCONA)* and showed how this tool simplifies security configuration management of services and policies in a system using risk measurement and mitigation. We also combine all the components into one single metric and present validation experiments using real-life vulnerability data from the *National Vulnerability Database (NVD)* and show a comparison with existing risk measurement tools.

INTRODUCTION

We can visualize a *computer network system* from both a physical and a security perspective. From a physical perspective, such a *system* can be considered as consisting of a number of computers, servers, and other system components interconnected via high-speed LAN or WAN. However, when we visualize a *system* from a security perspective, we can divide it into its service and network part. The network part allows data to come in to the system, some of which may be generated with the intent of an attack and are malicious in nature and

Autonomic Network Management Principles. DOI: 10.1016/B978-0-12-382190-4-00009-7

some are benign for the system. After such malicious data makes its way into the system, its impact depends on which services or softwares in the system are affected. Therefore, following this notion, a system can be considered as a combination of multiple services that interact with other outside systems using its network communication.

Since, we can model a system in such a way, risk evaluation of individual services can help in identifying services that pose higher risk. This in turn allows extra attention to be focused on such services. We can then incorporate this information in our network security policy or access options and enhance security. In this chapter, security policy is defined as a set of objectives, rules of behavior for users and administrators, list of active services, corresponding software, and finally, requirements for system configuration and management. Therefore, it also specifies which services are part of the system. But changes to security policy can only happen if we have a security measure of all the services. If the risk of a single service can be quantified, then existing aggregating methods can be used to evaluate the security of all the services in a system. Along with the service risk measurement, integrating it with the network/policy risk measurement can provide a complete risk analysis of the system. This can allow management to employ resources for risk mitigation based on the severity and importance of each of these two aspects of risk.

The effectiveness of a security metric depends on the security measurement techniques and tools that enable network administrators to analyze and evaluate network security. Our proposed new framework and user-friendly implementation for network security evaluation can quantitatively measure the security of a network based on these two critical risk aspects: the risk of having a successful attack due to the services and the risk of this attack being propagated within the network due to the policy. Finally all these components are integrated into a single metric called *Quality of Protection Metric (QoPM)* to portray the overall security level of the network for the decision makers.

We have a number of contributions in this chapter. First, our proposed tool, based on our framework, can help in comparing services and security policies with each other to determine which ones are more secure. It is also possible to judge the effect of a change to the policy by comparing the security metrics before and after the change. And this comparison can be done, in aggregated form (e.g., weighted average) or component wise, based on user needs. Second, our framework and its implementation represent an important step toward adaptive security systems in which networks can be evaluated and automatically hardened accordingly by constantly monitoring changes in the network and service vulnerabilities. We used the Java Programming Language to implement the proposed metrics in one graphical user-interface called *ROCONA*. This tool simplifies periodic risk measurement and management. Third, the underlying framework of our tool is more advanced with respect to the existing tools and previous research works. Previous work includes those that scan or analyze a given network to find out allowed traffic patterns and vulnerabilities in the services. Also, most of them do not predict the future state of the security of the network, or consider the policy immunity or spurious risk. Works that do try to predict future vulnerabilities such as [1] is limited to predicting software bugs, and it requires inside knowledge of the studied software. Our prediction model is general and can work using only publicly available data. Finally, using the National Vulnerability Database published by NIST [2], we performed extensive evaluation of our proposed model. The results show a high level of accuracy in the evaluation and a new direction toward security measurement.

This chapter first discusses related works, followed by a detailed description of the entire framework. Next, we present our strategy of combining all the scores; our implementation of the framework, deployment, and a case study; our experiments, results, and discussion on those results; and finally, our conclusion and some directions of future works.

RELATED WORK

Measurement of network and system security has always been an important aspect of research. Keeping this in mind, many organizational standards have been formulated to evaluate the security of an organization. The National Security Agency's (NSA) INFOSEC Evaluation Methodology (IEM) is an example in this respect. Details regarding the methodology can be found in [3]. In [4] NIST provides guidance to measure and strengthen the security through the development and use of metrics, but their guideline is focused on the individual organizations and they do not provide any general scheme for quality evaluation of a policy. There are some professional organizations as well as vulnerability assessment tools including Nessus, NRAT, Retina, and Bastille [5]. They actually try to find out vulnerabilities from the configuration information of the concerned network. However, all these approaches usually provide a report describing what should be done to keep the organization secured, and they do not consider the vulnerability history of the deployed services or policy structure.

A lot of research has been done on security policy evaluation and verification as well. Evaluation of VPN, Firewall, and Firewall security policies include [6, 8]. Attack graphs is another technique that has been well developed to assess the risks associated with network exploits. The implementations normally require intimate knowledge of the steps of attacks to be analyzed for every host in the network [9, 10]. In [11] the authors provide a way to do so even when the information is not complete. Still, this setup causes the modeling and analysis using this model to be highly complex and costly. Although Mehta et al. try to rank the states of an attack graph in [12], their work does not give any sort of prediction of future risks associated with the system, nor does it consider the policy resistance of firewall and IDS. Our technique that is employed to assess the policy's immunity to attack propagation implements a simpler and less computationally expensive technique that is more suitable for our goal.

Some research has been done on focusing on the attack surface of a network. Mandhata et al. in [13] have tried to find the attack surface from perspective of the attackability of system. Another work based on attack surface has been done by Howard et al. in [14] along channel, methods and data dimensions and their contributions in an attack. A. Atzeni et al. present a generic overall framework for network security evaluation in [15] and discuss the importance of security metrics in [16]. In [17] Pamula et al. propose a security metric based on the weakest adversary (i.e., the least amount of effort required to make an attack successful). In [1], Alhazmi et al. present their results on the prediction of vulnerabilities, and they report their work on the vulnerability discovery process in [18]. They argue that vulnerabilities are essentially defects in released software, and they use past data and the track records of the development team, code size, records of similar software, and the like to predict the number of vulnerabilities. Our work is more general in this respect and utilizes publicly available data.

Some research work has also focused on hardening the network. Wang et al. use attack graphs for this purpose in [19]. They also attempt to predict future alerts in multistep attacks using attack graphs [20]. A previous work on hardening the network was done by Noel et al. [21] in which they made use of dependencies among exploits. They use the graphs to find some initial conditions that, when disabled, will achieve the purpose of hardening the network. Sahinoglu et al. propose a framework in [22, 23] for calculating existing risk depending on present vulnerabilities in terms of threat represented as probability of exploiting this vulnerability and the lack of countermeasures. But all this work does not give us the total picture inasmuch as they predominantly try to find existing risk and fail to address how risky the system will be in the near future or how policy structure would impact on security. Their analysis regarding security policies cannot therefore be regarded as complete, and they lack the flexibility in evaluating them.

A preliminary investigation of measuring the existing vulnerability and some historical trends has been analyzed in a previous work [24]. That work was limited in analysis and scope, however.

SECURITY RISK EVALUATION FRAMEWORK

As can be seen in Figure 9.1, we have modeled our framework as a combination of two parts. The first part measures the security level of the services within the computer network system based on vulnerability analysis. This analysis considers the presence of existing vulnerabilities, the dormant risk based on previous history of vulnerabilities, and potential future risk. In the second part, the service exposure to outside network by filtering policies, the degree of penetration or impact of successful attacks against the network and the risk due to traffic destined for unused address space are measured from a network policy perspective. Service risk components, the exposure factor, and the unused address space exposure together give us the threat likelihood, and at the same

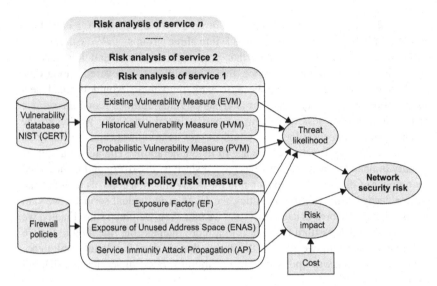

FIGURE 9.1   Network Risk Measurement Framework.

time the attack propagation provides us with the risk impact to the network of interest. When the risk impact is combined with the cost of the damage, we get the contribution of network policies in the total risk. In this chapter, we use the terms *vulnerability* and *risk* interchangeably. If a service has vulnerabilities, we consider it as posing a risk to the corresponding system. Also, the terms *security* and *risk* are regarded as opposites. Therefore, when we consider security, high values indicate a high level of security, and low values indicate low security. On the other hand, when we use the term *vulnerability*, high values indicate high vulnerability or risk (i.e., low security) and vice versa. Also, we refer to a whole *computer network system* (with both its service and network part) as simply a *system*.

Service Risk Analysis

In this section, we describe and discuss in detail the calculation method of our vulnerability analysis for service risk measurement. This analysis comprises *Historical Vulnerability Measure, Probabilistic Vulnerability Measure, and Existing Vulnerability Measure.*

Historical Vulnerability Measure

Using the vulnerability history of a service, the *Historical Vulnerability Measure (HVM)* measures how vulnerable a given service has been in the past. Considering both the frequency and age of the vulnerabilities, we combine the severity scores of past vulnerabilities so that a service with a high frequency of vulnerabilities in the near past has a high *HVM*.

Let the set of services in the system be S. We divide the set of vulnerabilities $HV(s_i)$ of each service $s_i \in S$ into three groups—$HV_H(s_i)$, $HV_M(s_i)$ and $HV_L(s_i)$—for vulnerabilities that pose high, medium, and low risks to the system. In evaluating a service, the vulnerabilities discovered a long time ago should carry less weight because with time these would be analyzed, understood, and patched. And our analysis of vulnerabilities found the service vulnerability to be less dependent on past vulnerabilities. So, we regard the relationship between vulnerability and its age as nonlinear, and we apply an exponential decay function of the age of the vulnerability. The age of a vulnerability indicates how much time has passed since its discovery and is measured in number of days. The parameter β in the decay function controls how fast the factor decays with age. In computing the HVM of individual services, we sum up the decayed scores in each class, and we take their weighted sum. Finally, we take its natural logarithm to bring it to a more manageable magnitude. Before taking the logarithm, we add 1 to the sum so that a sum of 0 will not result in ∞, and the result is always positive. The equation for HVM of service s_i, $HVM(s_i)$ is as follows:

$$HVM(s_i) = \ln\left(1 + \sum_{X \in \{H,M,L\}} w_X \cdot \sum_{v_j \in HV_X(s_i)} SS(v_j) \cdot e^{-\beta Age(v_j)}\right) \qquad (1)$$

In this equation, v_j is a vulnerability of service s_i, $Age(v_j)$ can have a range of $(0, \infty)$, and $SS(v_j)$ is its severity. Finally, the *Aggregate Historical Vulnerability Measure* of the whole system, *AHVM(S)*, is calculated as

$$AHVM(S) = \ln\left(\sum_{s_i \in S} e^{HVM}(s_i)\right) \qquad (2)$$

This equation is designed to be dominated by the highest HVM of the services exposed by the policy. We take the exponential average of all the HVM's so that the $AHVM$ score will be at least equal to the highest HVM and will increase with the HVM's of the other services. If arithmetic average was taken, then the risk of the most vulnerable services would have been undermined. Our formalization using the exponential average is validated through our conducted experiments.

Probabilistic Vulnerability Measure

The *Probabilistic Vulnerability Measure (PVM)* combines the probability of a vulnerability being discovered in the next period of time and the expected severity of that vulnerability. This measure, therefore, gives an indication of the risk faced by the system in the future.

Using the vulnerability history of a service, we can calculate the probability of at least one new vulnerability being published in a given period of time.

From the vulnerability history, we can also compute the expected severity of the predicted vulnerabilities.

We calculate *Expected Risk (ER)* for a service as the product of the probability of at least one new vulnerability affecting the service in the next period of time and the expected severity of the vulnerabilities. We can compare two services using this measure—a higher value of the measure will indicate a higher chance of that service being vulnerable in the near future.

For each service, we construct a list of the interarrival times between each pair of consecutive vulnerabilities published for that service. Then we compute the probability that the interarrival time is less than or equal to a given period of time, T. Let P_{s_i} be the probability that d_{s_i}, the number of days before the next vulnerability of the service s_i is exposed, is less than or equal to a given time interval, T, that is,

$$P_{s_i} = P(d_{s_i} \leq T) \tag{3}$$

To compute the expected severity, first we build the probability distribution of the severities of the vulnerabilities affecting the service in the past. Let X be the random variable corresponding to the severities, and $x_1, x_2, \ldots$ be the values taken by X. Then, the expectation of X_{s_i} for service S_i is given by Equation (4):

$$E[X_{s_i}] = \sum_{i=1, x_j \in s_i}^{\infty} x_j P(x_j) \tag{4}$$

where $P(x_j)$ is the probability that a vulnerability with severity x_j occurs for service s_i. Finally, we can define the expected risk, ER, of a service s_i as in Equation (5).

$$ER(s_i) = P_{s_i} \times E[X_{s_i}] \tag{5}$$

If S is the set of services exposed by the policy, we can combine the probabilities of all the services exposed by the policy to compute the *PVM* of the system as in Equation (6).

$$PVM(S) = \ln \sum_{s_i \in S} e^{ER(s_i)} \tag{6}$$

For combining the expected risks into one single measure, we are using the exponential average method like the *AHVM*, so that the *PVM* of a system is at least as high as the highest expected risk of a service in the system. Therefore, from a definition perspective, *PVM* is similar to *AHVM* as both measures correspond to a collection of services, whereas *HVM* and *ER* are similar as they correspond to a single service.

In order to calculate *PVM*, we have looked into different methods to model the interarrival times. The first method that we have analyzed is *Exponential Distribution*. Interarrival times usually follow exponential distribution [25].

In order to evaluate Equation (3) for a given service, we can fit the interarrival times to an exponential distribution, and we can find the required probability from the *Cumulative Distribution Function (CDF)*. If λ is the mean interarrival time of service s_i, then the interarrival times of service s_i, d_{s_i}, will be distributed exponentially with the following CDF:

$$P_{s_i} = P(d_{s_i} \leq T) = F_{d_{s_i}}(T) = 1 - e^{-\lambda T} \tag{7}$$

The next method that has been analyzed is *Empirical Distribution*. We can model the distribution of the interarrival times of the data using empirical distribution. The frequency distribution is used to construct a *CDF* for this purpose.

On the other hand, the empirical *CDF* of the interarrival time, d_{s_i}, will be:

$$P_{s_i} = P(d_{s_i} \leq T) = F_{d_{s_i}}(T) = \frac{\sum_{x \leq T} f_i(x)}{\sum f_i(x)} \tag{8}$$

Finally, exponential smoothing, a *Time Series Analysis* technique, was also used for identifying the underlying trend of the data and finding the probabilities.

Existing Vulnerability Measure

Existing vulnerability is important because sometimes the services in the system are left unpatched or in cases where there are no known patches available. Also, when a vulnerability is discovered, it takes time before a patch is introduced for it. During that time, the network and services are vulnerable to outside attack. The *Existing Vulnerability Measure (EVM)* measures this risk. *EVM* has been studied and formalized in our previous work [26].

We use the *exponential average* to quantify the worst case scenario so that the score is always at least as great as the maximum vulnerability value in the data. Let $EV(S)$ be the set of vulnerabilities that currently exist in the system with service set S, and let $SS(v_j)$ be the severity score of a vulnerability v_j. We divide $EV(S)$ into two sets—$EV_P(S)$ containing the vulnerabilities with existing solutions or patches, and $EV_U(S)$ containing those vulnerabilities that do not have existing solutions or patches. Mathematically,

$$EVM(S) = \alpha_1 \cdot \ln \sum_{v_j^p \in EV_P(S)} e^{SS(v_j^p)} + \alpha_2 \cdot \ln \sum_{v_j^u \in EV_U(S)} e^{SS(v_j^u)} \tag{9}$$

Here, the weight factors α_1 and α_2 are used to model the difference in security risks posed by those two classes of vulnerabilities. Details can be found in [26]. However, finding the risk to fully patched services based on historical trend and future prediction is one of the major challenges where we contribute in this chapter.

HVM-, PVM-, and EVM-based Risk Mitigation

There will always be some risks present in the system. But the vulnerability measures can allow the system administrator to take steps in mitigating system risk. As mentioned previously, *EVM* indicates whether there are existing vulnerabilities present in the system of interest. The vulnerabilities that may be present in the system can be divided into two categories. One category has known solutions, that is, patches available, and the other category does not have any solutions. In order to minimize risk present in the system, *ROCONA* (1) finds out which vulnerabilities have solutions and alerts the user with the list of patches available for install; (2) services having higher vulnerability measure than user defined threshold are listed to the user with the following options: (i) block the service completely, (ii) limit the traffic toward the service by inserting firewall rules, (iii) minimize traffic by inserting new rules in IDS, (iv) place the service in the DMZ area, and (v) Manual; (3) recalculates *EVM* to show the score for updated settings of the system. The administrator can set a low value as a threshold for keeping track of the services that contribute most in making the system vulnerable.

As can be seen from our definition of *HVM*, it gives us the historical profile of a software, and the *PVM* score gives us the latent risk toward the system. Naturally we would like to use a software that does not have a long history of vulnerabilities. In such cases a software with a low *HVM* and *PVM* score providing the same service should be preferred because low scores for them indicates low risk, as mentioned previously. Our proposed vulnerability calculation of *HVM* and *PVM* increases with the severity of vulnerabilities. Our implemented tool performs the following steps to mitigate the *HVM* and *PVM* risk. (1) Calculate the *HVM* and *PVM* scores of the services. (2) Compare the scores to the user-defined threshold values. (3) If scores are below the threshold, then strengthen the layered protection with options just as in the case of *EVM*. (4) Recalculate the *HVM* and *PVM* scores of the services. (5) If the scores are still below the threshold, then show recommendations to the system administrator. Recommendations include (i) isolation of the service using Virtual LANs, (ii) increase weight (i.e., importance) to the alerts originating from these services even if false alarms increase, and (iii) replace the service providing software with a different one.

The administrators can use our tool to measure the *EVM*, *HVM*, and *PVM* scores of a similar service providing software from the *NVD* database and choose the best possible solution. But for all of them, a cost-benefit analysis must precede this decision making. It should be clarified here that the three measures described here just provide us the risk of the system from a service perspective. If the network policy is so strong that no malicious data (i.e., attacks) can penetrate the system, then this score may not have high importance. But our assumption here is that a system will be connected to outside systems through the network and there will always be scope for malicious data to gain access within the system. That is why we use these vulnerability scores for decision making.

Network Risk Analysis

The network policies determine the exposure of the network to the outside world as well as the extensiveness of an attack on the network (i.e., how widespread the attack is). In order to quantify network risk, we have focused on three factors: *Exposure Factor (EF)*, *Attack Propagation (AP)*, and *Exposure of Unused Address Spaces (ENAS)*.

Attack Propagation Metric

The degree to which a policy allows an attack to spread within the system is given by the *Attack Propagation (AP)* factor. This measure assesses how difficult it is for an attacker to propagate an attack through a network across the system, using service vulnerabilities as well as security policy vulnerabilities.

Attack Immunity: For the purpose of these measures, we define a general system having N hosts and protected by k firewalls. Since we analyze this measure using a graph, we consider each host as a node in a graph where the graph indicates the whole system. The edges between nodes indicate network connectivity. Let each node be denoted as d and D be the set of nodes that are running at least one service. We also denote the set of services as S_d and p_{s_i} as the vulnerability score for service s_i in S_d. p_{s_i} has the range (0, 1).

For our analysis, we define a measure, I_{s_i}, that assesses the attack immunity of a given service, s_i, to vulnerabilities based on that service's *EVM, HVM,* and *PVM. I_{s_i}* is directly calculated from the combined vulnerability measure of a service s_i, P_{s_i} as:

$$I_{s_i} = -\ln(p_{s_i}) \tag{10}$$

I_{s_i} has a range of $[0, \infty)$, and will be used to measure the ease with which an attacker can propagate an attack from one host to the other using service s_i. Thus, if host m can connect to host n using service s_i exclusively, then I_{s_i} measures the immunity of host n to an attack initiating from host m. In case of a connection with multiple services, we define a measure of combined attack immunity. Before we can give the definition of the combined attack immunity, we need to provide one more definition. We define S_{mn} as the set of services that host m can use to connect to host n.

$$S_{mn} = \{s_i| \; host \; m \; connect \; to \; n \; using \; services \; s_i\} \tag{11}$$

Now, assuming that the combined vulnerability measure is independent for different services, we can calculate a combined vulnerability measure $p_{s_{mn}}$. Finally, using this measure, we can define the combined attack immunity $I_{s_{mn}}$ as:

$$I_{s_{mn}} = -\ln(p_{s_{mn}}) \times PL \tag{12}$$

Here *PL* is the protection level. For protection using a firewall, the protection is 1. For protection using *IDS*, the protection is a value between 0 and 1 and is equal to the false-negative rate of the *IDS*. It is assumed that the firewall will provide the highest immunity, and hence *PL* is assigned a value of 1. In case of *IDS*, the immunity decreases by its level of misses to attacks. This ensures that our *Attack Immunity* measure considers the absence of firewalls and *IDS*. From the definition of *AP*, it is obvious that high immunity indicates low risk to the system.

Service Connectivity Graph: To extract a measure of security for a given network, we map the network of interest to a *Service Connectivity Graph (SCG)*. The *SCG* is a directed graph that represents each host in the network by a vertex and represents each member of S_{mn} by a directed edge from *m* to *n* for each pair $m, n \in N$. Each arch between two hosts is labeled with the corresponding attack immunity measure. It should be noted that if an edge represents more than one service, then the weight of the edge corresponds to the combined attack immunity of the services connecting that pair of hosts. Considering that the number of services running on a host cannot exceed the number of ports (and is practically much lower than that), it is straightforward to find the time complexity of this algorithm to be $O(n^2)$. We assume that S_{mn} is calculated offline before the algorithm is run. A path through the network having a low combined historical immunity value represents a security issue. And if such a path exists, the system administrator needs to consider possible risk mitigation strategies focusing on that path.

Attack Propagation Calculation: For each node *d* in *D*, we want to find how difficult it is for an attacker to compromise all the hosts within reach from *d*. To do that, we build a minimum spanning tree for *d* for its segment of the *SCG*. The weight of this minimum spanning tree will represent how vulnerable this segment is to an attack. The more vulnerable the system is, the higher this weight will be. There are several algorithms for finding a minimum spanning tree in a directed graph [27] running in $O(n^2)$ time in the worst case. We define *Service Breach Effect (SBE$_d$)* to be the weight of the tree rooted at *d* in *D*. It actually denotes the damage possible through *d*. We use the following equation to calculate the *SBE$_d$*:

$$SBE_d = \sum_{n \in T} \left(\prod_{m \in \text{ nodes from } d \text{ to } n} p_{s_{dn}} \right) \times Cost_n \qquad (13)$$

Here *Cost$_n$* indicates the cost of damage when host *n* is compromised and *T* is the set of hosts present in the spanning tree rooted at host *d*. Finally, the attack propagation metric of the network is:

$$AP(D) = \sum_{d \in D} P(d) \times SBE_d \qquad (14)$$

where, $P(d)$ denotes the probability of the existence of a vulnerability in host d. If the service s_d present in host d is patched, then this probability can be derived directly from the expected risk $ER(s_d)$ defined previously in Equation (5). If the service is unpatched and there are existing vulnerabilities in service s_d of host d, then the value of $P(d)$ is 1. The equation is formulated such that it provides us with the expected cost as a result of attack propagation within the network.

Exposure of Unused Address Spaces (ENAS)

Policies should not allow spurious traffic, that is, traffic destined to *unused* IP addresses or port numbers, to flow inside the network because this spurious traffic consumes bandwidth and increases the possibility of a DDoS attack.[1] In our previous work [6, 7], we show how to identify automatically the rules in the security policy that allow for spurious traffic, and once we identify the possible spurious flows, we can readily compute the spurious residual risk for the policy.

Although spurious traffic may not exploit vulnerabilities, it can potentially cause flooding and DDoS attacks and therefore poses a security threat to the network. In order to accurately estimate the risk of the spurious traffic, we must consider how much spurious traffic can reach the internal network, the ratio of the spurious traffic to the total capacity, and the average available bandwidth used in each internal subnet. Assuming that maximum per-flow bandwidth allowed by the firewall policer is M_{l_i} Mbps for link l_i of capacity C_{l_i}, and F_{l_i}, is the set of all spurious flows passing through link l_i, we can estimate the *Residual Capacity (RC)* for link l_i as follows:

$$RC(l_i) = 1 - \frac{\sum_{f_j \in F_{l_i}} M_{l_i}}{C_{l_i}} \tag{15}$$

The residual capacity has a range [0, 1]. We can now calculate the *Spurious Risk (SPR)* of host d and then sum this measure for all the hosts in network A to get the *SPR* for the entire network:

$$SPR(A) = \sum_{d \in N} (c(d) \times \max_{l_i \in L} (1 - RC(l_i))) \tag{16}$$

where $c(d)$ is the weight (i.e., importance or cost) associated with the host d in the network with a range [0, 1] and L is the set of links connected to host d. We take the minimum of all the *Residual Capacity* associated with host d to reflect the maximum amount of spurious traffic entering the network and therefore measuring the worst case scenario. This is the spurious residual risk after considering the use of per-flow traffic policing as a countermeasure and assuming that each of the hosts within the network has a different cost value associated with each of them.

---

[1]We exclude here spurious traffic that might be forwarded only to Honeynets or sandboxes for analysis purposes.

Considering the Network Exposure Factor

In general, the risk of an attack may increase with the exposure of a service to the network. When we are using the severity score of a vulnerability, it should be multiplied by the EF to take into account the exposure of the service to the network. Therefore, this exposure influences EVM, HVM, and PVM calculations. We call it a *Network Exposure Factor (EF)*, and it is calculated as the fraction of the total address space to which a service is exposed. A service that is exposed to the total possible address space (e.g.,$*.*.*.*/ANY$) has a much higher probability of being attacked and exploited than one that is exposed only to the address space of the internal network. The Exposure Factor, EF, of a service s_i considers the number of IP Addresses served by s_i, $IP\ (s_i)$, the number of ports served by S, $PORTS\ (s_i)$, the total number of IP addresses, 2^{32} and the total number of ports, 2^{16}. Thus the range of this factor will start from the minimum value of 1 for the service totally hidden from the network and will reach the maximum value of 2 for the service totally exposed to the whole of the address space. Mathematically,

$$EF(s_i) = 1 + \frac{\log_2(IP(s_i) \times PORTS(s_i))}{\log_2(2^{32} \times 2^{16})} \tag{17}$$

assuming that the risk presented by the outside network is distributed uniformly. Of course, if we can identify which part of the address space is more risky than the others, we can modify the numerator of Equation (17) to be a weighted sum. For the purpose of multiplying the vulnerability severity score with EF, we can redefine the severity score as $SS(v_i) = $ Severity Score of the vulnerability $v_i \times EF(S_{v_i})$, where S_{v_i} is the service affected by the vulnerability v_i. This redefinition allows us to incorporate the EF in all the equations, where the severity of a vulnerability is being used, that is, Equations (1, 5, and 9).

AP- and SPR-based Risk Mitigation

AP indicates the penetrability of the network. Therefore, if this metric has a high value, it indicates that the network should be partitioned to minimize communication within the network and attack propagation as well. The possible risk mitigation measures that can be taken include (1) network redesigning (introducing virtual LANs); (2) rearrangement of the network so that machines having equivalent risk are in the same partition; (3) increase of the number of enforcement points (firewalls, IDS); (4) increased security around the hot spots in the network' and (5) strengthened MAC sensitivity labels, DAC file permission sets, access control lists, roles, or user profiles. In case of SPR, a high score for this metric indicates that the firewall rules and policies require fine tuning. When the score of this measure rises beyond the threshold level, $ROCONA$ adds additional firewall rules to the firewall. The allowable IP addresses and ports need to be checked thoroughly so that no IP address or port is allowed unintentionally.

QUALITY OF PROTECTION METRIC

Although the main goal of this research is to devise new factors in measuring network security, we show here that these factors can be used as a vector or a scalar element for evaluating or comparing system security or risk. For a system A, we can combine $EVM(A)$, $AHVM(A)$, $PVM(A)$, $AP(A)$, and $SPR(A)$ into one Total Vulnerability Measure, $TVM(A)$, as a vector containing $EVM(A)$, $AHVM(A)$, $PVM(A)$, $AP(A)$ and $SPR(A)$, as in Equation (18).

$$\mathbf{TVM}(A) = [EVM\,(A)\,AHVM\,(A)\,PVM\,(A)\,AP\,(A)\,SPR\,(A)]^{T} \qquad (18)$$

$$\mathbf{QoPM}(A) = 10\left[e^{-\gamma_1 EVM(A)}\,e^{-\gamma_2 AHVM(A)}\,e^{-\gamma_3 PVM(A)}\,e^{-\gamma_4(10-AP(A))}\,e^{-\gamma_5 SPR(A)}\right]^{T} \qquad (19)$$

This will assign the value of $[10\ 10\ 10\ 10\ 10]^{T}$ to a system with **TVM** of $[0\ 0\ 0\ 0\ 0]^{T}$. The component values assigned by this equation will be monotonically decreasing functions of the components of the total vulnerability measure of the system. The parameters γ_1, γ_2, γ_3, γ_4, and γ_5 provide control over how fast the components of the Policy Evaluation Metric decrease with the risk factors. The $QoPM$ can be converted from a vector value to a scalar value by a suitable transformation like taking the norm or using weighted averaging. For comparing two systems, we can compare their $QoPM$ vectors componentwise

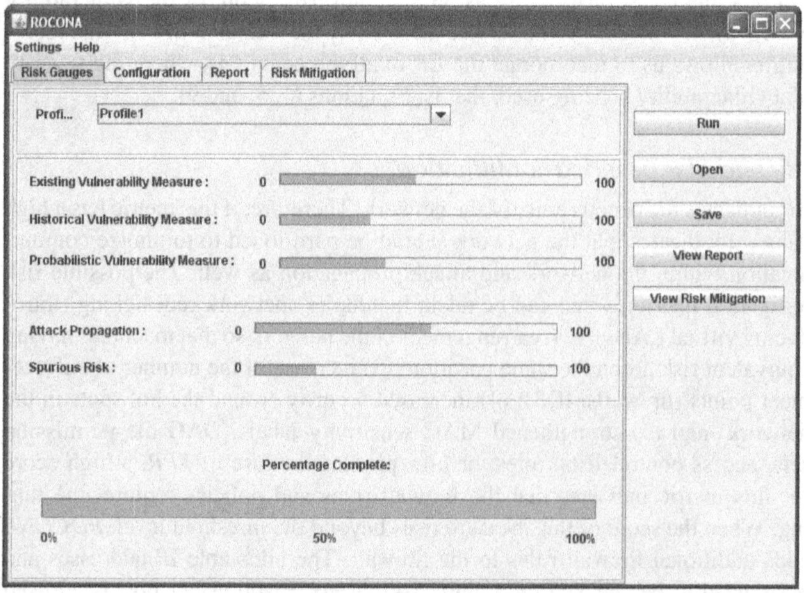

FIGURE 9.2   ROCONA Risk Gauges.

or convert the metrics to scalar values and compare the scalars. Intuitively, one way to combine these factors is to choose the maximum risk factor as the dominating one (e.g., vulnerability risk vs. spurious risk) and then measure the impact of this risk using the policy resistance measure. Although we advocate generating a combined metric, we also believe that this combination framework should be customizable to accommodate user preferences such as benefit sensitivity or mission criticality, which, however, is beyond the scope of this book. Another important aspect of this score is that, for the vulnerability measures (i.e., *EVM*, *HVM*, *PVM*), a higher score indicates higher risk or low security, whereas for *QoPM*, it is the opposite. In case of *QoPM*, a higher score indicates a higher level of security or lower risk toward the system.

ROCONA TOOL IMPLEMENTATION

To simplify the network risk measurement and mitigation using our proposed framework, we have implemented the measures in a tool called *Risk-based Proactive Security Configuration Manager (ROCONA)*. We have used the Java Programming Language for this implementation. The tool can run as a daemon process and therefore provide system administrators with periodic risk updates. The risk scores for each of the five components have been provided. The measures are provided as risk gauges. The risk is shown as a value between 0 and 100 as can be seen from Figure 9.2.

The users can set up different profiles for each run of the risk measurement. They can also configure different options such as parameters for *EVM*, *HVM*, and *PVM*, as well as the topology and vulnerability database files to consider during vulnerability measurement. Such parameters include the following.

1. The α_1 and α_2 weights required for measurement of *EVM*.
2. The decay factor β in the measurement of *HVM*.
3. The *Prediction Interval Parameter (T)* in *PVM* measurement.
4. The network topology file describing the interconnections between the nodes within the network.
5. The .xml files containing the list of vulnerabilities from NVD.
6. Option for manually providing the list of CVE vulnerabilities present in the system or automatically extracting them from third-party vulnerability scanning software. The current version can only use Nessus for the automatic option.
7. The starting date from which vulnerabilities are considered (for example, all vulnerabilities from 01/01/2004).

All these configuration options are shown in Figure 9.3. After the tool completes its run, it provides the system administrator with the measurement process in details. All the details can be seen in the *Details* tab. Using all the gauges, *ROCONA* also provides risk mitigation strategies (i.e., which service needs to patched, how rigorous firewall policies should be, etc.), which have already

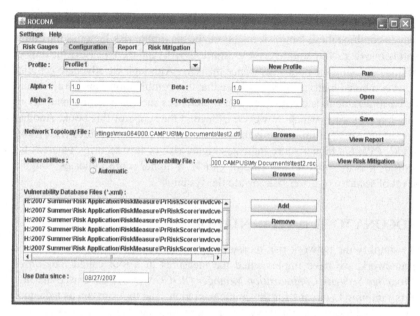

FIGURE 9.3 ROCONA Configuration.

been described with the individual measures in the *Risk Mitigation* tab. All this information can be saved for future reference or for comparison purposes.

Deployment and Case Study

The *ROCONA* tool has both a server and a client side. The client side is installed in individual computers in the system, whereas the server part is installed for the administrator of the system. Each client communicates with the server to get the parameter values provided by the administrator. Third-party software must also be installed so that the client program can extract the services currently running on a particular computer. The client side then sends the scores for individual components to the server where all the scores are combined to provide a unified score for the whole network system. We deployed our tool in two of our computer systems in *the University of Texas at Dallas Database Laboratory* to perform a comparative analysis of the risk for both the machines. Since both machines are in the same network and under the same firewall rules, the *AP* and *SPR* values have been ignored (i.e., they are equal). The comparison, therefore, includes *HVM*, *PVM*, and *EVM*. The results generated by *ROCONA* are then compared with the results of two well-known security tools: *AOL Active Security Monitor* [28] and *Nessus*. *AOL Active Security Monitor (ASM)* provides a score based on seven factors, which include factors such as firewall, anti-virus, spyware protection, and p2p software. *Nessus*, on the other hand, scans for open ports and how those ports can be used to compromise the system. Based on this,

TABLE 9.1 ROCONA Deployment and Evaluation

System	ROCONA			AOL Active Security Center	Nessus		
	HVM	PVM	EVM		High	Medium	Low
A	16.422	10.76	4.114	64	3	1	0
B	18.57	11.026	5.149	51	7	1	0

Nessus provides a report warning the user of the potential threats. Table 9.1 provides the comparison.

As can be seen from the table, System B is more vulnerable than System A. For System B, all of *HVM, PVM,* and *EVM* show higher values indicating higher security risk. The same trend can be seen using the other two tools as well. For *Active Security Monitor*, a higher value indicates lower risk, and for *Nessus*, the numbers indicate how many high, medium, or low vulnerabilities are in the system. We also performed comparisons between systems where both had the same service and software, but one was updated and the other was not. In that case only the *EVM* were found to be different (the updated system had lower *EVM*), and the *HVM* and *PVM* scores remained the same.

It may not be feasible to provide a side-by-side comparison of *ROCONA* and other tools because *ROCONA* offers unique features in analyzing risk, such as measuring the vulnerability history trend and prediction. However, this experiment shows that *ROCONA* analysis is consistent with the overlapping results of other vulnerability analysis tools such as *AOL Active Security Monitor* and *Nessus*. Another important feature of *ROCONA* is its dynamic nature. Since new vulnerabilities are found frequently, for the same state (i.e., the same services and network configuration) of a system, vulnerability scores will be different at two different times (if the two times are significantly apart) as new vulnerabilities may appear for the services in the system. But in such a case, *Nessus* and *AOL* would provide the same scores because their criteria of security measurement would remain unchanged. So far to our knowledge, there is no existing tool that can perform such dynamic risk measurement, and therefore, we find it sufficient to provide a comparison with these two tools. Here the scores themselves may not provide much information about the risk but can present a comparison over time regarding the system's state of security. This allows effective monitoring of the risk toward the system.

EXPERIMENTATION AND EVALUATION

We conducted extensive experiments for evaluating our metric using real vulnerability data. In contrast to similar research works that present studies on a few specific systems and products, we experimented using publicly available vulnerability databases. We evaluated and tested our metric, both componentwise and as a whole, on a large number of services and randomly generated policies. In our evaluation process, we divided the data into training sets and test sets. In the following sections, we describe our experiments and present their results.

Vulnerability Database Used In the Experiments

In our experiments, we used the National Vulnerability Database (NVD) published by National Institute of Science and Technology (NIST), which is available at http://nvd.nist.gov/download.cfm. The NVD provides a rich array of information that makes it the vulnerability database of choice. First of all, all the vulnerabilities are stored using the standard CVE (Common Vulnerabilities and Exposures) namehttp://cve.mitre.org/. For each vulnerability, the NVD provides the products and versions affected, descriptions, impacts, cross-references, solutions, loss types, vulnerability types, the severity class and score, and so on. The *NVD* severity score has a range of (0, 10). If the severity value is from 0 to 4, it is considered as low severity; scores up to 7 are considered medium severity; and a value higher than 7 indicates high severity. We have used the database snapshot updated at April 5, 2007. We present some summary statistics about the NVD database snapshot that we used in our experiments in Table 9.2.

For each vulnerability in the database, NVD provides CVSS scores [29] for vulnerabilities in the range 1 to 10. The severity score is calculated using the Common Vulnerability Scoring System (CVSS), which provides a base score depending on several factors such as impact, access complexity, and required authentication level.

Validation of HVM

We conducted an experiment to evaluate the *HVM* score according to Equation (1) using the hypothesis that if service A has a higher *HVM* than service B, then in the next period of time, service A will display a higher number of vulnerabilities than B.

TABLE 9.2 Summary Statistics of the NVD Vulnerability Database

Total number of entries	23542
Total number of valid entries	23309
Total number of rejected entries	233
Total number of distinct products	10375
Total number of versions of the products	147640
Total number of distinct vendors	6229
Earliest entry	10/01/1988
Latest entry	04/05/2007

In our experiment, we used vulnerability data up to June 30, 2006 to compute the *HVM* of the services, and we used the rest of the data to validate the result. We varied β so that the decay function falls to 0.1 in 0.5, 1, 1.5, and 2 years, respectively, and we observed the best accuracy in the first case. Here, we first chose services with at least 10 vulnerabilities in their lifetimes, and we gradually increased this lower limit to 100 and observed that the accuracy increases with the lower limit. As expected of a historical measure, better results have been found when more history is available for the services and observed the maximum accuracy of 83.33%. Figure 9.4(a) presents the results of this experiment.

Validation of Expected Risk (ER)

The experiment for validation of *Expected Risk (ER)* is divided into a number of parts. First, we conducted experiments to evaluate the different ways of calculating the probability in Equation (3). We conducted experiments to

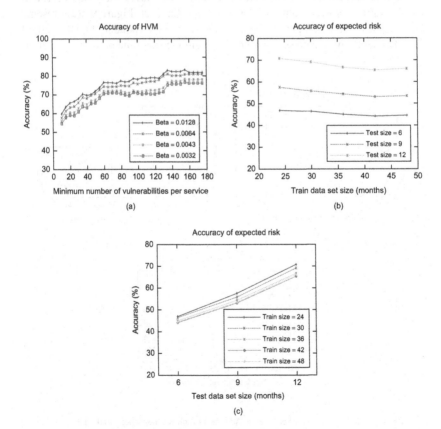

FIGURE 9.4 (a) Accuracy of the HVM for different values of β and minimum number of vulnerabilities. (b) Results of ER validation with training data set size vs. accuracy for different test data set sizes. (c) Results of ER validation with test data set size vs. accuracy for different training data set sizes.

compare the accuracies obtained by the exponential distribution, empirical distribution, and time series analysis method. Our general approach was to partition the data into training and test data sets, compute the quantities of interest from the training data sets, and validate the computed quantity using the test data sets. Here, we obtained the most accurate and stable results using exponential CDF.

The data used in the experiment for exponential CDF was the interarrival times for the vulnerability exposures for the services in the database. We varied the length of the training data and the test data set. We only considered those services that have at least 10 distinct vulnerability release dates in the 48-month training period. For Expected Severity, we used a similar approach. For evaluating expected risk, we combined the data sets for the probability calculation methods and the data sets of the expected severity.

In the experiment for exponential CDF, we constructed an exponential distribution for the interarrival time data and computed Equation (3) using the formula in Equation (7). For each training set, we varied the value of T and ran validation for each value of T with the test data set. Figure 9.5(a) presents the accuracy of the computed probabilities using the Exponential CDF method for different training data set sizes and different values of the prediction interval

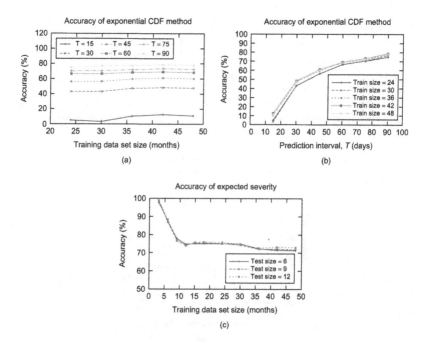

FIGURE 9.5   (a) Training Data Set Size vs. Accuracy graph for the Exponential CDF method for probability calculation, test set size = 12 months. (b) Prediction Interval Parameter (T) vs. Accuracy graph for the Exponential CDF method for probability calculation, test set size = 12 months. (c) Training data set size vs. Accuracy graph for the Expected Severity calculation for different values of the test data set size.

parameter, T. In Figure 9.5(b), we have presented the accuracies on the Y axis against the prediction interval parameter T on the Y axis. For the test data set size of 12 months, we observed the highest accuracy of 78.62% with the 95% confidence interval being [72.37, 84.87] for the training data set size of 42 months and prediction interval of 90 days. We present the results of the experiment for Expected Severity in Figure 9.5(c). The maximum accuracy obtained in this experiment was 98.94% for the training data set size of 3 months and the test data set size of 9 months. The results of the Expected Risk experiment are presented in the Figures 9.4(b) and 9.4(c). For the Expected Risk, we observed the best accuracy of 70.77% for training data set size of 24 months and test data set size of 12 months with the 95% confidence interval of [64.40, 77.14].

It can be observed in Figure 9.5(b) that the accuracy of the model increases with increasing values of the prediction interval parameter T. This implies that this method is not sensitive to the volume of training data available to it. From Figure 9.5(c), it is readily observable that the accuracy of the Expected Severity is not dependent on the test data set size. It increases quite sharply with decreasing values of training set data size for small values of training set data size. This means that the expectation calculated from the most recent data is actually the best model for the expected severity in the test data.

Validation of QoPM

To validate the $QoPM$, we need to evaluate policies using our proposed metric in Equation (19). In the absence of other comparable measure of system security, we used the following hypothesis: If system A has a better $QoPM$ than system B based on training period data, then system A will have a smaller number of vulnerabilities than system B in the test period. We assume that the EVM component of the measure will be 0 as any existing vulnerability can be removed.

In generating the data set, we chose the reference date separating the training and test periods to be October 16, 2005. We used the training data set for ER using the Exponential CDF method. In the experiment, we generated a set of random policies, and for each policy we evaluated Equation (19). We chose $\gamma_1 = \gamma_2 = \gamma_3 = \gamma_4 = \gamma_5 = 0.06931472$ so that the $QoPM$ for a TVM of [10 10 10 10 10]^T is [5 5 5 5 5]^T. Then, for each pair of systems A and B, we evaluated the hypothesis that if $\mathbf{QoPM}(A) \geq \mathbf{QoPM}(B)$, then the number of vulnerabilities for service A should be less than or equal to the number of vulnerabilities for service B. In our experiment, we varied the number of policies from 50 to 250 in steps of 25. In generating the policies, we varied the number of services per system from 2 to 20 in increments of 2.

We present the results obtained by the experiment in Figures 9.6(a) and 9.6(b). As mentioned previously, a policy can be regarded as a set of rules indicating which services are allowed access to the network traffic. We set up different service combinations and consider them as separate policies for our experiments. We can observe from the graph in Figures 9.6(a) that the accuracy

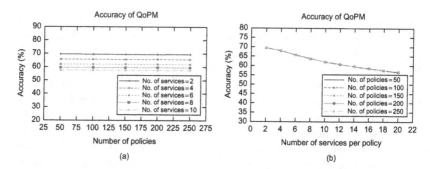

FIGURE 9.6 (a) Number of Policies vs. Accuracy Graph for QoPM for different values of the number of services in each policy. (b) Number of Services per Policy vs. Accuracy Graph for QoPM for different values of the number of polices.

of the model is not at all sensitive to the variation in the total number of policies. However, the accuracy does vary with the number of services per policy: The accuracy decreases with increasing number of services per policy. This trend is more clearly illustrated in Figures 9.6(b) where the negative correlation is clearly visible.

Running Time Evaluation of the Attack Propagation Metric

To assess the feasibility of calculating the AP metric under different conditions, we ran a Matlab implementation of the algorithm for different network sizes, as well as different levels of network connectivity percentages (the average percentage of the network directly reachable from a host). The machine used to run the simulation was a 1.9 GHz P-IV processor with 768 MB RAM. The results are shown in Figures 9.7. For each network size, we generated several random networks with the same % connectivity value. The running time was then calculated for several hosts within each network. The average value of the running time per host was then calculated and used in Figures 9.7. As can be seen from the figure, the quadratic growth of the running time against network connectivity was not noticeably dependent on the network %connectivity in our implementation. The highest running time per host for a network of 320 nodes was very reasonable at less than 5 seconds. Thus, the algorithm scales gracefully in practice and is thus feasible to run for a wide range of network sizes.

CONCLUSIONS

As network security is of the utmost importance for an organization, a unified policy evaluation metric will be highly effective in assessing the protection of the current policy and in justifying consequent decisions to strengthen security. In this chapter, we present a proactive approach to quantitatively evaluate

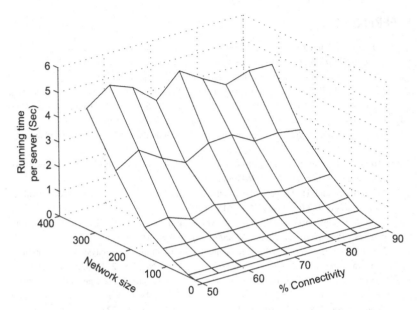

FIGURE 9.7   The Average Running Times for Different Network Sizes and % Connectivity.

security of network systems by identifying, formulating, and validating several important factors that greatly affect its security. Our experiments validate our hypothesis that if a service has a highly vulnerability-prone history, then there is a higher probability that the service will become vulnerable again in the near future. These metrics also indicate how the internal firewall policies affect the security of the network as a whole. These metrics are useful not only for administrators to evaluate policy/network changes and, take timely and judicious decisions, but also for enabling adaptive security systems based on vulnerability and network changes.

Our experiments provide very promising results regarding our metric. Our vulnerability prediction model proved to be up to 78% accurate, while the accuracy level of our historical vulnerability measurement was 83.33% based on real-life data from the *National Vulnerability Database (NVD)*. The accuracies obtained in these experiments vindicate our claims about the components of our metric and also the metric as a whole. Combining all the measures into a single metric and performing experiments based on this single metric also provides us with an idea of its effectiveness.

ACKNOWLEDGMENTS

The authors would like to thank M. Abedin and S. Nessa of the University of Texas at Dallas for their help with the formalization and experiments making this work possible.

REFERENCES

[1]   O.H. Alhazmi and Y.K. Malaiya, "Prediction capabilities of vulnerability discovery models," *Proc. Reliability and Maintainability Symposium*, January 2006, pp. 86–91.

[2]   "National institute of science and technology (nist)," http://nvd.nist.gov

[3]   R. Rogers, E. Fuller, G. Miles, M. Hoagberg, T. Schack, T. Dykstra, and B. Cunningham, *Network Security Evaluation Using the NSA IEM*, 1st ed., Syngress Publishing, August 2005.

[4]   M. Swanson, N. Bartol, J. Sabato, J. Hash, and L. Gra3o, *Security Metrics Guide for Information Technology Systems*. National Institute of Standards and Technology, Gaithersburg, MD, July 2003.

[5]   "10 network security assessment tools you can't live without," http://www.windowsitpro.com/Article/ ArticleID/47648/47648.html?Ad=1

[6]   E. Al-Shaer and H. Hamed, "Discovery of policy anomalies in distributed firewalls," *Proceedings of IEEE INFOCOM'04*, March 2004.

[7]   H. Hamed, E. Al-Shaer, and W. Marrero, "Modeling and verification of ipsec and vpn security policies," *Proceedings of IEEE ICNP'2005*, November 2005.

[8]   S. Kamara, S. Fahmy, E. Schultz, F. Kerschbaum, and M. Frantzen, "Analysis of vulnerabilities in internet firewalls," *Computers and Security*, Vol. 22, No. 3, p. 214–232, April 2003.

[9]   P. Ammann, D. Wijesekera, and S. Kaushik, "Scalable, graph-based network vulnerability analysis," in *CCS '02: Proceedings of the 9th ACM conference on Computer and Communications Security*. New York, ACM Press, 2002, pp. 217–224.

[10] C. Phillips and L.P. Swiler, "A graph-based system for network-vulnerability analysis," *NSPW '98: Proceedings of the 1998 Workshop on New Security Paradigms*. New York, ACM Press, 1998, pp. 71–79.

[11] C. Feng and S. Jin-Shu, "A flexible approach to measuring network security using attack graphs," *International Symposium on Electronic Commerce and Security*, April 2008.

[12] C.B. Mehta, H. Zhu, E. Clarke, and J. Wing, "Ranking attack graphs," *Recent Advances in Intrusion Detection 2006*, Hamburg, Germany, September 2006.

[13] P. Manadhata and J. Wing, "An attack surface metric," *First Workshop on Security Metrics*, Vancouver, BC, August 2006.

[14] M. Howard, J. Pincus, and J.M. Wing, "Measuring relative attack surfaces," *Workshop on Advanced Developments in Software and Systems Security*, Taipei, December 2003.

[15] A. Atzeni, A. Lioy, and L. Tamburino, "A generic overall framework for network security evaluation," *Congresso Annuale AICA 2005*, October 2005, pp. 605–615.

[16] A. Atzeni and A. Lioy, "Why to adopt a security metric? a little survey," *QoP-2005: Quality of Protection Workshop*, September 2005.

[17] J. Pamula, P. Ammann, S. Jajodia, and V. Swarup, "A weakest-adversary security metric for network configuration security analysis," *ACM 2nd Workshop on Quality of Protection 2006*, Alexandria, VA, October 2006.

[18] O.H. Alhazmi and Y.K. Malaiya, "Modeling the vulnerability discovery process," *Proc. International Symposium on Software Reliability Engineering*, November 2005.

[19] L. Wang, S. Noel, and S. Jajodia, "Minimum-cost network hardening using attack graphs," *Computer Communications,* Alexandria, VA, November 2006.

[20] L. Wang, A. Liu, and S. Jajodia, "Using attack graphs for correlating, hypothesizing, and predicting intrusion alerts," *Computer Communications*, September 2006.

[21] S. Noel, S. Jajodia, B. O'Berry, and M. Jacobs, "Efficient minimum-cost network hardening via exploit dependency graphs," *19th Annual Computer Security Applications Conference*, Las Vegas, NV, December 2003.

[22] M. Sahinoglu, "Security meter: A practical decision-tree model to quantify risk," *IEEE Security and Privacy*, June 2005.

[23] M. Sahinoglu, "Quantitative risk assessment for dependent vulnerabilities," *Reliability and Maintainability Symposium*, June 2005.

[24] M.S. Ahmed, E. Al-Shaer, and L. Khan, "A novel quantitative approach for measuring network security," *INFOCOM'08*, April 2008.

[25] S.C. Lee and L.B. Davis, "Learning from experience: Operating system vulnerability trends," *IT Professional*, Vol. 5, No. 1, January/February 2003.

[26] M. Abedin, S. Nessa, E. Al-Shaer, and L. Khan, "Vulnerability analysis for evaluating quality of protection of security policies," *2nd ACM CCS Workshop on Quality of Protection*, Alexandria, VA, October 2006.

[27] F. Bock, "An algorithm to construct a minimum directed spanning tree in a directed network," *Developments in Operations Research*, Gordon and Breach, 1971, pp. 29–44.

[28] "Aol software to improve pc security," http://www.timewarner.com/corp/newsroom/pr/ 0,20812,1201969,00. html

[29] M. Schi3man, "A complete guide to the common vulnerability scoring system (cvss)," http://www.first.org/ cvss/cvss-guide.html, June 2005.

Chapter 10

The Design of the FOCALE Autonomic Networking Architecture

John Strassner

ABSTRACT

Network devices will always be heterogeneous, both in the functionality they provide and in the way they represent and use management data. This adversely affects interoperability and makes management of networks and networked applications more difficult. This chapter describes the motivation and design of the FOCALE autonomic networking architecture. FOCALE is based on the following core principles: (1) use a combination of information and data models to establish a common "lingua franca" to map vendor- and technology-specific functionality to a common platform-, technology-, and language-independent form, (2) augment this with ontologies to attach formally defined meaning and semantics to the facts defined in the models, (3) use the combination of models and ontologies to discover and program semantically similar functionality for heterogeneous devices independent of the data and language used by each device, (4) use context-aware policy management to govern the resources and services provided, (5) use multiple-control loops to provide adaptive control to changing context, and (6) use multiple machine learning algorithms to enable FOCALE to be aware of both itself and of its environment in order to reduce the amount of work required by human administrators. This chapter first motivates the need for autonomic systems and explains why a well-known but simple example of an autonomic control loop is not sufficient for network management purposes. It uses these deficiencies as motivation to explain the rationale behind the original FOCALE autonomic architecture. The chapter concludes with a discussion of how knowledge is represented in FOCALE.

INTRODUCTION AND BACKGROUND

Network administration is inherently difficult. In contrast to other types of administration, network administration typically involves manually managing diverse technologies and devices whose functionality needs to be coordinated to provide a set of services to multiple users. This creates complex situations

Autonomic Network Management Principles DOI: 10.1016/B978-0-12-382190-4.00010-3 that are prone to many different types of errors, due to the byzantine relationships between end users, the needs of their applications, and the capabilities of network devices that support those applications. The automation of complex, manually intensive configuration tasks that are prone to error is a primary motivation for autonomic systems. This section will describe the motivation for building the FOCALE autonomic architecture [1]. FOCALE stands for Foundation–Observe–Compare–Act–Learn–rEason, which describes its novel operation in terms of the structure of its unique control loops.

This section will introduce the concept of autonomic systems and explain why they have great promise for solving some of most difficult problems in network management.

Current Network Management Problems

Existing Business Support Systems (BSSs) and Operational Support Systems (OSSs) are designed in a *stovepipe* fashion [2]. This enables the OSS or BSS integrator to use systems and components that are best suited to their specific management approach. However, this approach hampers interoperability because each management component uses its own view of the managed environment, thereby creating information islands. An example of an industry OSS is shown in Figure 10.1. This particular OSS exemplifies the underlying *complexity* of integrating the many different subsystems in a modern OSS.

Referring to Figure 10.1, three important problems are exhibited:

- Connections are *brittle*: Changing any one component affects a set of others that are directly connected, which in turns affects a new set of components, and so on . . .

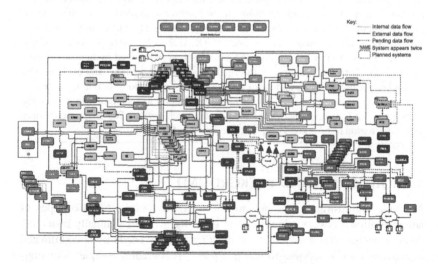

FIGURE 10.1   An Example of an Operational Support System.

- The learning curve for understanding this architecture, let alone maintaining and upgrading pieces of it, is enormous.
- The lack of commonly defined data prohibits different components from sharing and reusing common data.

It is easily seen that these problems can *not* be solved by new graphical user interfaces (GUIs) or protocols. The use of different GUIs simply complicates matters, since a GUI inherently shows *its* own application-centric view of the portion of the network that it is managing, and *not* that of the entire system. The use of different protocols doesn't help because different protocols have different limitations in the data that they can query for, the commands that they can transmit, and the data that they can represent.

Different Forms of Complexity

The above three problems are all different forms of *complexity*. Legacy approaches have tried to solve this problem by building a set of application-specific data models that represent the characteristics of devices and management applications. This is shown in Figure 10.2.

In the above OSS, the inventory solution is shown as being designed in a way that facilitates the operation of the inventory solution. While this enables it to represent data in the best way for it to operate, this approach is *not* the best way for the system as a whole to use this data. This is because many different applications all share a view of the same customer. Unfortunately, if they are designed in disparate ways without regard to each other, then the same data is likely to be represented in different formats. In Figure 10.2, the Inventory

FIGURE 10.2 Infrastructural Shortcomings Resulting from Stovepipe Architectures.

Management and the Configuration Management applications name the same customer in different ways. Furthermore, while both use the concept of an ID, one calls it "CUSTID" and the other calls it simply "ID." In addition, the former represents its ID as an integer datatype, while the latter uses a string datatype. These and other problems make the finding, let alone sharing, of information more difficult. This problem is exacerbated by the need to use multiple specialized applications, and also because of the large number of distinct types of applications that are required.

This simple example illustrates the tendency of each application to build its own view of the data in its own way. This creates stovepipes and makes it very hard (if not impossible) for data to be shared. Note that the above was a very simplistic example. In general, a *mapping* (which can be quite complicated, as it often involves functions that transform the data) is required to associate the same or related data from different applications, since they have varying attribute names and values having different attributes in different formats to each other. This results in a set of architectural and integration issues listed in Figure 10.2.

The Inherent Heterogeneity of Management Data

Unfortunately, even if all OSS and BSS application vendors could agree on a common way to represent information, there is another source of complexity: the programming model of the device. In general, each network device has its own programming model or models. The Internet Engineering Task Force (IETF) has published a standard called the Structure of Management Information (SMI) [3], which is the grammar for writing a type of data model, called a Management Information Base (MIB), for managing devices in a communications network. Unfortunately, this standard has important fundamental problems that compromise its utility.

Figure 10.3 illustrates one of the problems with this approach. An MIB contains managed objects that are organized in a hierarchy. Entries in the tree are addressed using object identifiers, some of which are shown in Figure 10.3. The problem is shown in the ellipse at the bottom—the "private" subtree is used to contain vendor-specific information, while other branches are used to contain interoperable public data. Unfortunately, most of the important information required to manage a device is private information. Hence, even though SMI is a standard, accessed through the Simple Network Management Protocol (SNMP, which is another set of standards) [4], the lack of standard data that different vendors *must* use inhibits its utility.

There are additional problems with SNMP and SMI that are beyond the scope of this chapter; the interested reader is encouraged to read [5] for a good overview of these. Due to these and other problems, it is common for operators to use vendor-specific Command Line Interface (CLI) [6] programming models for provisioning. If the operator wants to use SNMP for monitoring, then another important problem is introduced: *There is no underlying common*

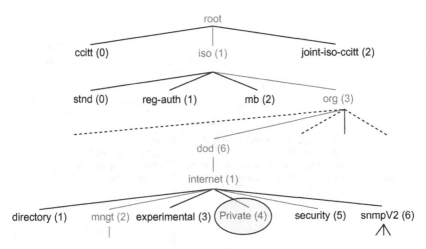

FIGURE 10.3 MIB Naming (Simplified).

information model that relates CLI commands to SNMP commands. Without this common model, it is impossible to build a closed loop system because the administrator doesn't know what SNMP commands to issue to check if a set of CLI commands acted as expected.

Figure 10.4 shows an example of the complexity caused by different programming models of network devices—a system administrator defining two different routers from two different manufacturers as BGP peers. Some of the important problems that appear include the following.

- The syntax of the language used by each vendor is different.
- The semantics of the language used by each vendor is different.
- The router on the left uses *modes*, which are not present in the router on the right.
- The router on the right supports virtual routing (via its routing-instances command), which the router on the left does not support.

There are other differences as well in their functionality that are not shown. In effect, by using these two different routers, the operator has built two different stovepipes in its network, and administrators are forced to learn the vendor-specific differences of each router. This causes difficult problems in maintaining semiautomated configuration solutions, such as script-based solutions using PERL, TCL, or EXPECT, since they cannot easily accommodate syntax changes. Command signature changes, such as additional or reordered parameters, exacerbate the problem and commonly occur when different versions of a network device CLI are introduced. This is how the industry builds OSSs and BSSs today.

Therefore, if this cannot be automated (due to its complexity), how can it be solved? Certainly not manually, since besides being the source of errors,

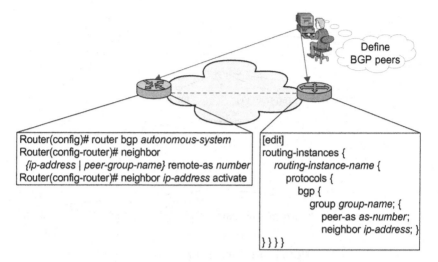

```
Router(config)# router bgp autonomous-system
Router(config-router)# neighbor
{ip-address | peer-group-name} remote-as number
Router(config-router)# neighbor ip-address activate
```

```
[edit]
routing-instances {
    routing-instance-name {
        protocols {
            bgp {
                group group-name; {
                    peer-as as-number;
                    neighbor ip-address; }
}}}}
```

FIGURE 10.4   Different Programming Models of Two Routers.

humans may not even be present, which means that problems can perpetuate and become worse in the system. To better understand this problem, consider the following analogy: The language spoken by a device corresponds to a language spoken by a person, and different dialects of a language correspond to different capabilities in different versions of the CLI. How can a single administrator, or even a small number of administrators, learn such a large number of different languages for managing the different types of network devices in a network? There are *hundreds* of versions of a single release of a network device operating system, and there are tens to hundreds of different releases. More importantly, if there is an end-to-end path through the network consisting of multiple devices, how can this path be configured and managed without a single "über-language"?

The Purpose of Autonomic Systems

The previous section has described some of the many complexities that are inherent in network management. One of the ways of dealing with these complexities is to use autonomic systems. The primary purpose of autonomic systems is to *manage complexity* [1, 2, 7]. The name was chosen to reflect the function of the autonomic nervous system in the human body. The analogy is not exact, as many of the decisions made by the autonomic nervous system of the human body are involuntary, whereas ideally, most decisions of autonomic systems have been delegated to a human-designed governance structure, such as policy-based management. However, it is indicative of the general nature of autonomics, which is to offload as many manually intensive tasks from users as possible in order to enable those users to spend their time more productively.

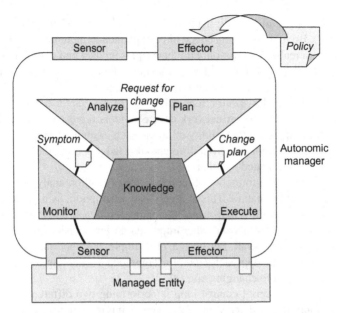

FIGURE 10.5 A Simplified View of IBM's Autonomic Control Loop.

The heart of an autonomic system is its ability to manage its operation by taking an appropriate set of actions based on the conditions it senses in the environment [2, 7]. Sensors retrieve data, which is then analyzed to determine if any correction to the managed resource(s) being monitored is needed (e.g., to correct nonoptimal, failed, or error states). If so, then those corrections are planned, and appropriate actions are executed using effectors that translate commands back to a form that the managed resource(s) can understand. If the autonomic network can perform manual, time-consuming tasks (such as configuration management) on behalf of the network administrator, then that will free the system and the administrator to work together to perform higher-level cognitive functions, such as planning and network optimization.

Figure 10.5 shows a high-level diagram, taken from [7], that depicts the simple control loop previously described. While this appears simple, it makes the following assumptions that are typically not valid for network environments. First, the entity being managed can be directly instrumented (e.g., with sensors that get data and effectors that issue commands). This is not true in most network devices, as the device provides a strict and unyielding interface and/or set of APIs that provide a view into data that it gathers and enables *some*, but not *all*, software functionality to be implemented. For example, most network equipment manufacturers will *not* provide direct control over the operating system used in the network device.

Second, networks consist of heterogeneous data and languages. It is common for a single network device to support both Command Line Interface (CLI) and management applications built using the Simple Network Management Protocol (SNMP). Both of these concepts will be explained in more detail; for now, it is sufficient to realize that not only do they represent data and commands completely differently for different devices, a single device can support *hundreds* of different versions of a particular CLI release, *tens* of different CLI releases, and *hundreds* of vendor-specific representations of data that are in general not interoperable with how other network device vendors represent the same data. This brings into question the validity of taking such a simple approach, where it is assumed that different data from different knowledge sources can be so easily represented and integrated together.

Third, note that the four control loop functions (monitor, analyze, plan, and execute) are *serial* in nature. This is a very bad assumption, as it prevents different parts of the control loop from interacting with each other. Section 3 in Chapter 11 will discuss this and other important design decisions of autonomic control loops in more detail. For now, a very simple, but important, counterexample exists that this approach *cannot* accommodate: enabling a governance mechanism to *adapt* to changing context.

Finally, a single serial control loop is performing two different functions. First, it is monitoring and analyzing data; second, it is reconfiguring one or more managed entities. How can a single control loop perform two functions that are completely different in purpose, method, algorithm, and most importantly, affected entities (e.g., to solve a problem, a device different from the one being monitored may need to be reconfigured, which means that the control loop starts and ends at different places!)?

A simplified block diagram of the original FOCALE autonomic architecture [1] is shown in Figure 10.6. A unique and novel feature of FOCALE is its use of *context-aware policy rules* to govern the functionality of the management system and to direct the operation and execution of its control loop elements.

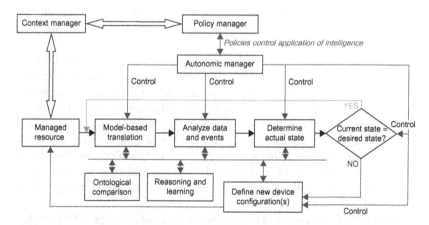

FIGURE 10.6 A Simplified View of the FOCALE Autonomic Architecture.

Networks can contain hundreds of thousands of policy rules of varying types (e.g., high-level business policy rules for determining the services and resources that are offered to a user, to low-level policy rules for controlling how the configuration of a device is changed; a good introduction to this is [8]). One of the purposes of making these policy rules context-aware is to use context to select only those policy rules that are applicable to the current management task being performed. Examples of such an approach are given in [8, 9]. The Policy Manager is responsible for translating business requirements written in a restricted natural language designed for readability and usability into a form that can be used to configure network resources [8–10]. For example, the business rule "John gets Gold Service" will be translated into a set of configuration commands that are written in a lower-level language (such as C or Java) or that can be translated into a set of CLI or SNMP commands (for direct application to a network device) that will be applied to all affected devices.

This translation is done via the Policy Continuum [8, 9, 11–13], which forms a continuum of policies in which each policy captures the requirements of a particular constituency. For example, businesspeople can define an SLA in business terms, focusing on revenue; this is then transformed into architectural requirements and finally network configuration commands. Note that each policy representation of the SLA does not "disappear" in this approach; rather, they function as an integrated whole (or continuum) to enable the translation of business requirements to network configuration commands. This is a crucial step in preserving the semantics of different policy rules, which would otherwise be lost if simple syntactic translations were used instead.

Different devices and networks having vastly different functionality, such as wired and wireless networks, can be interconnected. In general, in addition to the aforementioned differences in technology and programming models, each device has different semantics, and each network has different business goals; all of these can be captured by a combination of policy rules and context. This enables different policy rules to be used for the same device if the type of applications or environment changes. FOCALE solves this problem by using context-aware policy rules to vary the functionality of its control loop in order to address these needs, as well as changes in context, user needs, and business objectives. The autonomic manager of FOCALE dynamically adjusts the function of each element of the control loop, which is shown by the set of "control" arrows connecting the autonomic manager to each function in the control loop.

The reconfiguration process uses dynamic code generation [2, 9, 14]. Information and data models are used to populate the state machines that in turn specify the operation of each entity that the autonomic system is governing. The management information that the autonomic system is monitoring consists of captured sensor data. This is analyzed to derive the current state of the managed resource, as well as to alert the autonomic manager of any context changes in or involving the managed resource. The autonomic manager then compares the current state of the entities being managed to their desired state; if the states are equal, then monitoring continues. However, if the states are not equal, the autonomic manager will compute the optimal set of state transitions required to change the states of the entities being managed to their corresponding desired states. During this process, the system could encounter an unplanned change in context (e.g., if a policy rule that has executed produced undesirable side effects). Therefore, the system checks, as part of both the monitoring and configuration control loops, whether or not context has changed. If context has not changed, the process continues. However, if context has changed, then the system first adjusts the set of policies that are being used to govern the system according to the nature of the context changes, which in turn supplies new information to the state machines. The goal of the reconfiguration process is specified by state machines; hence, new configuration commands are dynamically constructed from these state machines. We automate simple, recurring reconfigurations using vendor-specific command templates.

The purpose of this short introduction has been to highlight two foundational problems that have often been overlooked in the design of autonomic systems: (1) the inability to share and reuse data from different sources, and (2) the inability to *understand the meaning* of sensor data, so that the system can react in as efficient (and correct) a manner as possible. The next section will examine how knowledge is represented, so that machine-based learning and reasoning can be performed. This enables FOCALE to combine different data from different sources and integrate them into a cohesive whole to better understand how the system is operating and how it may be improved.

REPRESENTING KNOWLEDGE

One of the first problems to solve in building an autonomic architecture for network management is the inherent heterogeneity present in the devices and technologies used. Just as there is still no one dominant programming language, there will likely never be a single dominant language for representing management and operational data and commands. More importantly, there is another important consideration. Network equipment manufacturers have spent many years—even decades—building and fine-tuning their languages, whether they are CLI- or SNMP-based. Thus, even if a single "über-language" was built, there is no *business motivation* for these vendors to *retool* their products to use such a language. Therefore, the approach taken in FOCALE is to accept that vendor-specific languages, along with vendor-specific representations of management and operational data, will remain. FOCALE therefore uses a single language *internally* to simplify processing and *maps* different vendor-specific management and operational data into this internal form. Similarly, when remediation commands are computed, FOCALE maps these technology- and vendor-neutral commands back into a vendor-specific form. This requires a common way of representing knowledge. In FOCALE, this is done using a combination of information and data models augmented with ontologies.

The Role of Information and Data Models in FOCALE

Since vendors will always use one or more proprietary languages to represent management and operational data, a proven approach to solving this heterogeneity problem is to use this single information model to define the management information definitions and representations that *all* OSS and BSS components will use [2]. This is shown in Figure 10.7. The top layer represents a single, enterprisewide information model, from which multiple standards-based data models are derived (e.g., one for relational databases and one for directories). Since most vendors add their own extensions, the lowest layer provides a second model mapping to build a high-performance vendor-specific implementation (e.g., translate SQL92 standard commands to a vendor proprietary version of SQL). Note that this is especially important when an object containing multiple attributes is split up, so that some of the object's attributes are in one data store and the rest of the object's attributes are in a different data store. Without this set of hierarchical relationships, data consistency and coherency are lost.

The idea behind Figure 10.7 is to design a mediation layer that can use a single information model to obtain the basic characteristics and behavior of managed entities in a technology-, language-, and platform-independent manner. This enables each existing vendor-specific application, with its existing stovepipe data model, to be mapped to the above data model, enabling applications to share and reuse data. In the BGP peering example shown in Figure 10.4, a single information model representing common functionality (peering using BGP) is defined; software can then be built that enables this single common information model to be translated to vendor-specific implementations that support the different languages of each router manufacturer. Hence, the complexity of representing information is reduced from n^2 (where each vendor's language is mapped to every other language) to n mappings (to a common data model).

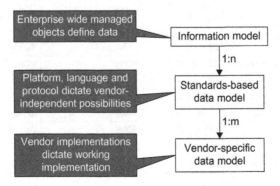

FIGURE 10.7   The Relationship between Information Models and Data Models.

Choosing an Information Model

The Unified Modeling Language (UML) [15] is the de facto standard for building information and data models. DEN-ng [16] is the next generation of the original DEN [17] standard. It is a comprehensive object-oriented UML-based information model that uses software patterns [18] to model managed entities. It has been built in a modular and structured way to facilitate close linkage with a set of ontologies, which can be used to overcome UML limitations through their use of formal logic.

DEN-ng is significantly different from other object-oriented models, such as the DMTF's Common Information Model (CIM) [19] and the TMF's Shared Information and Data (SID) [20] model. While the SID stays true to the original goals of DEN, the CIM does not. Part of the reason is that the CIM is in reality a *data model*, as it uses database concepts such as keys and weak relationships as a fundamental part of its model; this violates the mandate that an information model is independent of platform, language, and protocol. A full comparison is given in [16] and partly summarized in Figure 10.8.

While all of the above features are important in building an autonomic architecture, the fact that the SID and CIM do not have any notion of policy application and negotiation, state machine, context, metadata, and ontology compatibility means that they do not provide enough technical features to satisfy the needs of autonomic architectures. Hence, FOCALE has chosen the DEN-ng model, both for its extensibility and for the advanced functionality that it offers.

Feature	DEN-ng	SID	CIM
True information model?	YES	YES	NO – uses database concepts
Classification theory	Strictly used	Not consistently used; not used by mTOP	Not used at all
Patterns	Many more used that the SID	4	Not used at all
Views	Life-chain, not life-cycle	Business only; mTOP defining informal system & impl views	One view
Policy model	DEN-ng v7	DEN-ng v3.5	Simple IETF model
Policy application model	DEN-ng v7	NONE	NONE
Policy negotiation model	DEN-ng v6.7	NONE	NONE
State machine model	DEN-ng v6.8	NONE	NONE
Ontology compatibility	STRONG	NONE	NONE
Capability model	DEN-ng v6.6.4	NONE	Very simplified
Context model	DEN-ng v6.6.4	NONE	NONE
Metadata model	DEN-ng v7	NONE	NONE

FIGURE 10.8 A Simplified Comparison of the DEN-ng, CIM, and SID Models.

Organizing Knowledge Using the DEN-ng Information Model

DEN-ng defines a single root class that has three subclasses: Entity, Value, and MetaData. This is shown in Figure 10.9, with only a few exemplary attributes shown to keep the diagram simple. (Note that both the CIM and the SID have a large number of subclasses of their root class, which complicates the usage and understanding of their models.)

The knowledge representation approach in FOCALE is based on federating knowledge extracted from different sources. This enables an application to use different knowledge representation mechanisms for different types of data. The different sets of knowledge are then fused into a final knowledge representation that the application can use. This is accommodated by the use of classification theory [21, 22] to create structurally pure class hierarchies that are logically consistent and contain a minimum of "exceptions." This enables data to be differentiated from information and information from knowledge. In fact, the set of processes that transform data into information, and then into knowledge, are critical for defining and representing knowledge in a reusable, scalable way. Data is characterized as observable and possibly measurable raw values that signal something of interest. In and of themselves, data have no meaning—they are simply raw values. In contrast, data is transformed into information when meaning can be attached to data. This provides the data with a simple, starting context. The process of transforming information into knowledge attaches purpose, forms a more complete context, and provides the potential to generate action.

However, there is a profound difference between modeling a fact that is observed or measured and modeling a fact that is inferred. Facts that are

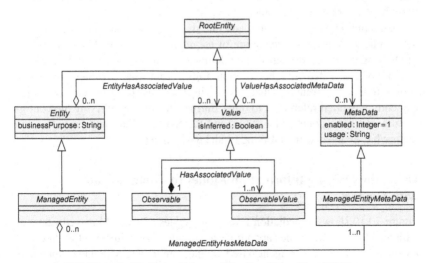

FIGURE 10.9 The Entity, Value, and MetaData Hierarchies of DEN-ng.

observed or measured often do not need additional reasoning performed on them. For example, if the speed of a device interface is measured to be 100 Megabits per second, then that measurement is all that is needed to define the fact. In stark contrast, facts that are inferred can only exist by having reasoning performed to create them. For example, a measured value of 17 is simply a scalar. If that value can be associated with a set of enumerated values to give the value 17 a particular meaning (e.g., "problem"), the system has now succeeded in attaching a meaning to the scalar value. If the system can add further details, such as what the problem refers to and what a possible solution to this problem could be, the semantics are now made more explicit, and important additional information and knowledge can now be generated (i.e., a skill set could be inferred as required to solve the problem whose value was denoted as 17). This systematic enrichment of semantics, from a simple value to a value with meaning to a value with meaning, purpose, and action, is critical to defining knowledge that can be used to perform a task. This is reflected in the three top-level hierarchies that are represented by the Entity, Value, and MetaData subclasses of the DEN-ng information model in Figure 10.9 and in differentiating measured values from inferred values in appropriate subclasses.

A unique feature of DEN-ng is that it models the *aspects* of a managed entity; other models represent a managed entity as an atomic object. This enables DEN-ng to (1) build models that are made up of reusable building blocks that represent specific aspects of an overall task, and (2) clearly separate the concept of managed entities via the roles that they take on. For example, DEN-ng uses the role object pattern [23] (and variations of it) to model roles for both humans as well as nonhuman entities, such as locations (this area allows unsecured Internet access), logical concepts (device interfaces that communicate with access devices and networks compared to those that communicate with the network core), and other concepts.

A unique feature of DEN-ng, and one that is very important for managing autonomic systems, is the modeling of *metadata*. The rich metadata hierarchy of DEN-ng increases flexibility in modeling concepts by separating characteristics and behavior that are inherent in an Entity from ways in which that Entity is used (e.g., the current role that it is playing). The use of metadata also avoids embedding explicit references to other Entities, thereby simplifying the configuration and provisioning of Entities and the Services that they provide. This feature has evolved as the DEN-ng model has evolved.

Using the DEN-ng Information Model to Communicate with Devices

Figure 10.10 shows a mediation layer that enables a legacy Managed Element with no inherent autonomic capabilities to communicate with an autonomic system. This is realized as an agent-based architecture in FOCALE and is known as the model-based translation layer (MBTL). DEN-ng supplies the "model" part

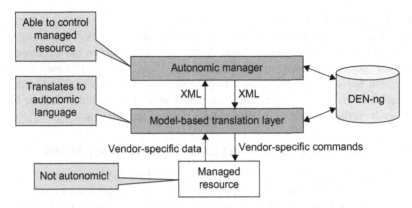

FIGURE 10.10 The Model-Based Translation Layer of FOCALE.

of the translation layer, enabling the Managed Element to communicate with other autonomic elements.

The MBTL accepts vendor-specific data and translates it into a normalized representation of that data, described in XML, for further processing by the autonomic manager. This provides a common language for representing input sensor data. The agents themselves are built using the factory pattern [24] and consist of parsers to translate the input data into a normalized XML document. The DEN-ng model is used to supply data used by the parsers. The factory pattern is recursive; hence, repeated application of the factory pattern enables libraries to be built that handle the languages supported by a given device as a set of factories that call other lower-level factories as needed.

Once the autonomic manager has completed analyzing the data, it may need to issue one or more vendor-specific commands to correct any problems encountered. This is done by translating the (vendor-neutral) XML document into a set of vendor-specific commands to effect the reconfiguration.

The DEN-ng model plays a prominent role in the implementation of the MBTL. In FOCALE, state machines are used to orchestrate behavior. The management and operational data acquired from the sensors is matched to expected values from the appropriate data models, which are derived from the DEN-ng information model. This makes accommodating the requirements of heterogeneous devices and programming models much easier, as the information model shows how different data from different devices can be related to each other. These data are used to determine the current states of the entities being managed. This is compared to their desired states, and if they are not equal, appropriate reconfiguration commands are issued. In order to determine which commands to issue, each node in the state machine can be viewed as a set of configuration commands. Hence, determining the correct commands to issue corresponds to determining the optimal state transitions that are required to move the current states of the entities being managed to their respective desired states. The

DEN-ng information model contains a generic description of software that can be associated with devices, services, and applications (including the configuration of software commands); work in the FOCALE program has developed vendor-specific software models for different CLIs and SNMP-based management applications. Thus, the set of commands to be executed is broadly determined by the finite state machines, and in a more granular fashion by lower-level data models of the affected devices and applications.

Governance Using the DEN-ng Context-Aware Policy Model

A simplified extract of the DEN-ng context model is shown in Figure 10.11. In this figure, the Context class represents a complete representation of context, while the ContextData class represents a specific aspect of Context. For example, if the context of a user is downloading streaming multimedia, then this context can be represented as the aggregation of several different sets of contextual information, including time, location, type of connection, applicable business rules, and types of applications used. If these context aspects are each represented as independent objects that can be combined, then the management system can reuse known patterns to quickly and efficiently model context. Both Context and ContextData have associated role classes (only the latter is shown in the figure), just as other important managed entities, such as resources and services, do. In addition, policy rules, people, organizations, and other important entities have roles defined to govern the functionality that they can provide. In each case, DEN-ng uses the role-object pattern (or a variant of it) to standardize how roles represent functionality played by the entity having the roles.

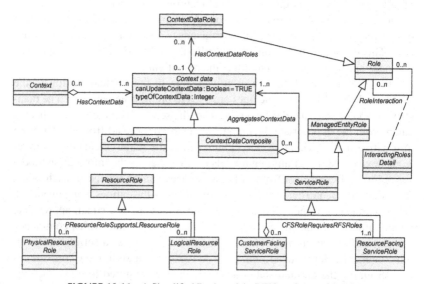

FIGURE 10.11   A Simplified Portion of the DEN-ng Context Model.

In this approach, a context role selects a set of applicable policy rules (or their roles), which in turn select the set of authorized roles that can be used for that context. These roles define the functionality that is allowed for this context. For network devices, this includes the services and resources that are offered; for the control loop, this is used to determine the characteristics and behavior of that particular control element. For example, it can be used to change the function of a control element, such as changing a comparison operation from a simple threshold comparison to a more comprehensive deep object comparison. Hence, when context changes, new policy rules are selected that change the functionality of the system to correspond with the context change. This is a unique feature in autonomic control loops.

Figure 10.12 shows the high-level design, along with a few key attributes, of the DEN-ng policy model. Note the use of the abstract class PolicyRuleStructure, which serves as a common root of three different types of policy rules. By way of comparison, the SID only defines ECA, or Event–Condition–Action, policy rules, and the CIM only defines CA, or Condition-Action, policy rules; neither the SID nor the CIM has the inherent extensibility to define alternate types of policy rules without performing a major redesign on their current policy rule models.

The PolicyRuleStructure class is used to define the concept of representing the structure of a policy rule. This enables the structure of a policy rule to be separated from the content of the policy rule, providing additional extensibility. The currently supported rule types include CA (condition–action, for backwards compatibility), ECA (event–condition–action, preferred over CA), Goal, and Utility policies. Common attributes and relationships that are independent of the particular representation of a policy rule are defined in the PolicyRuleStructure class. For example, Figure 10.12 defines two common attributes for handling policy rules that fail to execute correctly and, more importantly, define an aggregation that relates a set of PolicyRules to a set of State Machines. Thus, any type of DEN-ng PolicyRule will be able to use a state machine to govern behavior.

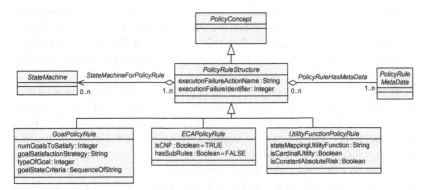

FIGURE 10.12   A Simplified Portion of the DEN-ng Policy Rule Model.

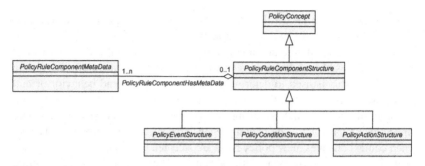

FIGURE 10.13   A Simplified Portion of the DEN-ng Policy Rule Component Model.

Subclasses of this class formalize the semantics of different types of Policy Rules using a subsumption relationship. This enables DEN-ng to import different types of Policy Rules, each with its own specific structure, and represent how each is used. This provides extensibility, so that new Policy Rule types can be added without adversely affecting the overall design of the DEN-ng Policy Hierarchy.

From an ontological perspective, it is important to separate the semantics of the structural representation of the policy rule from other concepts that are required to use the Policy Rule, such as Policy Target and Policy Subject. This enables other policy types, for example, ManagementPolicy, to define the appropriate semantics as required. An example of a subclass that can be defined from PolicyRuleStructure is ECAPolicyRule, which formalizes the semantics of a Policy Rule with an {Event, Condition, Action} structure. In essence, an ECAPolicyRule is a Policy Rule that has at least one of a PolicyEvent, a PolicyCondition, and a PolicyAction.

A unique feature of this model is the definition of the PolicyRuleMetaData class, which defines the basic metadata that applies to different types of policy rules, such as ECAPolicies. This decouples common metadata that different Policy representation systems need from the actual realization of the Policy, enabling all PolicyRules to share common metadata while enabling specific types of PolicyRules to define their own particular metadata. It also decouples the representation and structure of a particular type of policy (e.g., an ECAPolicy) from the metadata. This is critical for properly constructing ontologies from policy models.

A similar approach is used to for PolicyRuleComponents and is shown in Figure 10.13. The three subclasses repeat this abstraction in order to define different types of events, conditions, and actions; PolicyRuleComponentMetaData and PolicyRuleComponentStructure mirror the use of the PolicyRuleMetaData and PolicyRuleStructure classes.

Figure 10.14 shows conceptually how context is used to affect policy. The SelectsPoliciesToActivate aggregation defines a set of Policies that should be loaded and activated based on the current context. Hence, as context changes, policy can change accordingly, enabling our system to adapt to changing

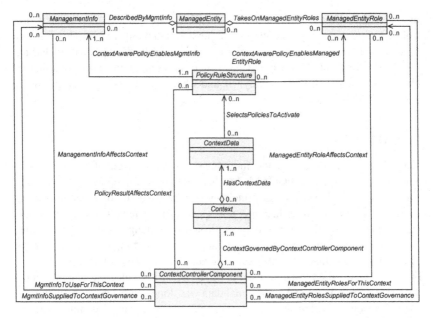

FIGURE 10.14   A Simplified View of the DEN-ng Context-Aware Policy Rule Model.

demands. Note that this selection is an "intelligent decision" in that the selection process depends on other components that are part of a particular context. Another "intelligent decision" is the PolicyResultAffectsContext association, which enables policy results to influence Context via the ContextController-Component, the application that manages Context. For example, if a policy execution fails, not only did the desired state change not occur, but the context may have changed as well.

The selected working set of Policies uses the ContextAwarePolicyEnables-ManagedEntityRoles association to define and enable the appropriate ManagedEntity roles that are influenced by this Context; each ManagedEntityRole defines functionality that the ManagedEntity can use. In this way, policy indirectly (through the use of roles) controls the functionality of the system, again as a function of context. Similarly, ContextAwarePolicyEnablesMgmtInfo defines the set of management data that is useful for this Context; ManagementInfoAffectsContext represents feedback from these management data regarding the execution of the policy rule. Once the management information is defined, then the two associations MgmtInfoAffectsContext and ManagedEntityRoleAffectsContext codify these dependencies (e.g., context defines the management information to monitor, and the values of these management data affect context, respectively).

Finally, the ContextControllerComponent defines its own set of ManagedEntityRoles and ManagementInfo to use to monitor the environment; feedback from the ManagedEntities identified by their roles and specific measurements (in the form of ManagementInfo) are used by the ContextControllerComponent to operate its own finite state machine (FSM). This FSM is used to orchestrate the actions of the ContextControllerComponent, including which ManagedEntities it should determine the context for, how a particular element of context should be analyzed, and what procedures for determining its context should be used.

SUMMARY

This chapter has described the motivation for and design of the FOCALE autonomic architecture. FOCALE has been designed for managing heterogeneous networks. However, its principles can be used for other applications, as the mechanisms described are not specifically oriented toward network management. In particular, the most difficult challenge that FOCALE has overcome—integrating heterogeneous management and operational data from different sources that use different languages to describe their data—makes FOCALE well suited to being used in other industries, since the integration of disparate management and operational data was done using a novel new knowledge representation that was independent of any underlying technology being managed.

This chapter has emphasized the importance of knowledge representation and fusion. Too often, network management focuses on a specific technology or device, or a set of technologies and devices, and ignores the larger picture of how network services and resources are used by different users, as well as how environmental conditions affect the network services and resources offered. The future will increasingly be dominated by the need to enable services that have conflicting requirements of shared resources to coexist. In addition, context awareness—the ability for the resources and services offered at any given time to *change* as a function of user needs, environmental conditions, business goals, or other factors—must be taken into account. Static governance approaches that use static knowledge bases cannot deal with these complexities. Finally, nextgeneration and future Internet approaches are now emphasizing the *economic viability* of such architectures. That is, any future architecture must be economically sustainable. Therefore, an important requirement of next-generation and future architectures is to be able to *use business demands to drive the offering of services and resources at any particular time.*

This latter approach is clearly beyond the capabilities of current management languages, such as CLI and SNMP. Such a new knowledge representation and language will most likely *not* come from a network equipment vendor because there is no *business reason* for a network equipment vendor to *retool* its existing device operating system and languages to provide such a language.

Knowledge representation is a critical portion of an autonomic system because all autonomic systems make decisions based on sensor data. One of the principal distinguishing features of FOCALE is that it doesn't just retrieve sensor data, it attempts to *understand* the significance of sensor data. This enables long-term action, such as predicting faults based on current performance data trends. In addition, it is impossible to provide context-aware functionality without understanding the significance of data and correlating data values to the current context. Indeed, the same data may mean different things based on context; hence, it is imperative to associate increased semantics with data to enable machine learning and reasoning to be performed.

The FOCALE approach recommends integrating models and ontologies. Models provide a powerful way to describe facts and support advanced code generation methodologies. Ontologies represent a powerful set of tools to attach semantic meaning to data. The FOCALE architecture is a specific architecture that uses models and ontologies to represent knowledge. The Model-Based Translation module converts vendor-specific sensor data into a common vendor-neutral form, and vendor-independent commands into vendor-specific versions. The autonomic manager determines the state of the entity being managed from received sensor data and/or inferences and determines what actions to take guided by context-aware policy rules. Semantic reasoning helps orchestrate system behavior through the use of machine-based learning (which enables system operation to be improved with experience) and reasoning (which generates hypotheses of root causes of problems and proves whether or not a hypothesis is correct).

REFERENCES

[1] J. Strassner, N. Agoulmine, and E. Lehtihet, "FOCALE—A novel autonomic networking architecture," *International Transactions on Systems, Science, and Applications (ITSSA) Journal*, Vol. 3, No. 1, pp. 64–79, May 2007.

[2] J. Strassner, "Management of autonomic systems—Theory and practice," Network Operations and Management Symposium (NOMS) 2010 Tutorial, Osaka, Japan, April 2010.

[3] SMI is defined by a set of RFCs, including RFC2576.txt, RFC2578.txt, RFC2579.txt, and RFC2580.txt., These can be downloaded using the following form: http://www.mibdepot.com/downloads/rfcs/rfc25xy.txt.

[4] D. Harrington, R. Preshun, and B. Wijnen, "An architecture for describing simple network management protocol management frameworks," RFC3411, STD0062, December 2002.

[5] J. Schönwälder, A. Pras, and J.-P. Martin-Flatin, "On the future of Internet management technologies," *IEEE Communications Magazine*, Vol. 41, No. 10, pp. 90–97, October 2003.

[6] See, for example, http://www.cisco.com/warp/cpropub/45/tutorial.htm.

[7] IBM, "An architectural blueprint for autonomic computing," June 2006, 4th ed., available for download from: http://www-01.ibm.com/software/tivoli/autonomic/pdfs/AC_Blueprint_White_Paper_4th.pdf.

[8] J. Strassner, *Policy-Based Network Management*, Morgan Kaufman Publishers, September 2003.

[9] J. Strassner, "Enabling autonomic network management decisions using a novel semantic representation and reasoning approach," Ph.D. thesis, 2008.

[10] J. Strassner, S. van der Meer, D. O'Sullivan, and S. Dobson, "The use of context-aware policies and ontologies to facilitate business-aware network management," *Journal of Network and System Management*, Vol. 17, No. 3, pp. 255–284, September 2009.

[11] J. Strassner, J.N. de Souza, S. van der Meer, S. Davy, K. Barrett, D. Raymer, and S. Samu-drala, "The design of a new policy model to support ontology-driven reasoning for autonomic networking," *Journal of Network and Systems Management*, Vol. 17, No. 1, pp. 5–32, March 2009.

[12] S. Davy, B. Jennings, and J. Strassner, "The policy continuum—Policy authoring and conflict analysis," *Computer Communications Journal*, Vol. 31, No. 13, pp. 2981–2995, August 2008.

[13] S. Davy, B. Jennings, and J. Strassner, "The policy continuum—A formal model," in Proc. of the 2nd International IEEE Workshop on Modeling Autonomic Communications Environments (MACE), eds. B. Jennings, J. Serrat and J. Strassner, Multicon Lecture Notes— No. 6, Multicon, Berlin, pp. 65–78.

[14] J. Strassner and D. Raymer, "Implementing next generation services using policy-based management and autonomic computing principles," Network Operations and Management Symposium, pp. 1–15, Vancouver, Canada, 2006.

[15] The set of UML specifications, defined by the Object Management Group, are here: www.uml.org.

[16] J. Strassner, "Introduction to DEN-ng", Tutorial for FP7 PanLab II Project, January 21, 2009, available from: http://www.autonomic-communication.org/teaching/ais/slides/0809/Introduction_to_DEN-ng_for_PII.pdf.

[17] J. Strassner, *Directory Enabled Networks*, Macmillan Technical Publishing, 1999.

[18] E. Gamma, R. Helm, R. Johnson, and J. Vlissides, *Design Patterns: Elements of Reusable Object-Oriented Software*, Addison-Wesley, 1994.

[19] http://www.dmtf.org/standards/cim.

[20] http://www.tmforum.org/InformationFramework/1684/home.html.

[21] J. Parsons, "An information model based on classification theory," *Management Science*, Vol. 42, No. 10, pp. 1437–1453, October 1996.

[22] J. Parsons and Y. Wand, "Emancipating instances from the tyranny of classes in information modeling," *ACM Transactions on Database Systems*, Vol. 25, No. 2, pp. 228–268, June 2000.

[23] The role object pattern is available for download from: http://st-www.cs.uiuc.edu/users/hanmer/PLoP-97/Proceedings/riehle.pdf

[24] The factory pattern may be downloaded from: www.patterndepot.com/put/8/factory.pdf.

Chapter 11

Knowledge Representation, Processing, and Governance in the FOCALE Autonomic Architecture

John Strassner

ABSTRACT

The previous chapter has established the motivation for building the FOCALE autonomic architecture. This chapter traces the evolution of FOCALE by first explaining why its knowledge representation is based on the integration of knowledge extracted from information and data models with knowledge extracted from ontologies. It then explains the development of the novel FOCALE control loops, which use its unique knowledge representation to support an advanced cognition model that is based on cognitive psychology. This cognition model provides FOCALE with a unique feature that other autonomic architectures do not have: the ability to emulate how human beings make decisions.

INTRODUCTION AND BACKGROUND

The previous chapter described the FOCALE autonomic architecture, which is based on the following core principles: (1) use a combination of information and data models to establish a common "lingua franca" to map vendor- and technology-specific functionality to a common platform-, technology-, and language-independent form, (2) augment this with ontologies to attach formally defined meaning and semantics to the facts defined in the models, (3) use the combination of models and ontologies to discover and program semantically similar functionality for heterogeneous devices independent of the data and language used by each device, (4) use context-aware policy management to govern the resources and services provided, (5) use multiple control loops to provide adaptive control to changing context, and (6) use multiple machine learning algorithms to enable FOCALE to be aware of both itself and of its environment in order to reduce the amount of work required by human administrators.

Autonomic Network Management Principles. DOI: 10.1016/B978-0-12-382190-4.00011-5

While the first version of FOCALE was very successful, it was not able to *dynamically adapt* quickly enough to changing context. In addition, it was not able to make use of advanced decision-making algorithms. For example, if it detected a change in context, it was not able to adapt its control loops quickly enough to enable a smooth transition to the new context. This chapter explores how the FOCALE architecture evolved to solve these problems.

KNOWLEDGE PROCESSING IN FOCALE

Current network management approaches encode fixed knowledge to manage networks. For example, a service provider aggregates application traffic into a small number of classes and then defines a set of traffic conditioning policy rules that control the classification, marking, dropping, queuing, and scheduling behavior of traffic for a network for these specific traffic classes. These policy rules are installed into the network. This approach works well as long as the traffic characteristics and usage are as expected. However, if the network conditions are not as expected, either because the users generate different traffic than expected or because of other environmental conditions, then this approach will not be able to dynamically adjust to changing user needs, business goals, and environmental conditions.

In contrast, the goal of FOCALE is to dynamically adjust the services and resources offered in accordance with such changes. This means that FOCALE requires a *dynamically updateable* knowledge base. FOCALE supports this requirement by using semantic reasoning to examine sensed management and operational data not just to analyze its current values, but more importantly, to determine whether there are any patterns between different data, or whether expected results are different from those expected. This latter feature is especially significant, for it enables the system to verify and validate data against its knowledge base and to update the knowledge base when new information is discovered. This is in general not possible with the vast majority of current network management systems, since they rely on static knowledge. FOCALE performs the validation and verification of management and operational data by using semantic reasoning, which employs first-order logic to reason about the validity of the new or changed information; if it can be proved with reasonable certainty that this information is correct, then it is added to the knowledge base. This enables the FOCALE knowledge base to evolve with experience.

The knowledge approach used in FOCALE is described in detail in [1] and is based on federating domain-specific knowledge extracted from different sources; these sources can be information and/or data models as well as ontologies. The algorithm used can optionally weight the contribution of each data source to construct an overall understanding of the operation of a managed system with a given environment according to the contributions and/or importance of each source. The knowledge that collectively describes the behavior of an entity or process is then fused into a final integrated form that the application can use.

Why UML-Based Models Are Insufficient to Represent Network Knowledge

While UML is an excellent tool, its lack of formality has given rise to a number of serious errors in its metamodel [2]. More importantly, the metamodel lacks the ability to represent fundamental pieces of knowledge, such as:

- Relationships,[1] such as "is similar to," which are needed to relate different vocabularies (such as commands from different vendors) to each other
- formal semantics of sets
- lexical relationships, such as synonyms, antonyms, homonyms, and meronyms
- ability to link many representations of an entity to each other (e.g., precise and imprecise)
- groups whose components are *not* individually identifiable, only quantifiable
- richer sentence semantics, through representations of assertions, queries, and sentences about sentences as *objects*

Furthermore, while UML is itself a language, it does not have the ability to model the languages used in network devices. This is one reason network management and administration is still so difficult—as soon as an operator uses more than one vendor and more than one version of a language, there is an immediate jump in complexity for building some means to translate between the different languages that are used to manage devices. The above points all focus on the *inability to share and reuse knowledge due to the lack of a common language*. This has been pointed out as far back as 1991, where it was stated that with respect to representing and acquiring knowledge, "Knowledge-acquisition tools and current development methodologies will not make this problem go away because the root of the problem is that knowledge is inherently complex and the task of capturing it is correspondingly complex. ...Building qualitatively bigger knowledge-based systems will be possible only when we are able to share our knowledge and build on each other's labor and experience." [3]

Finally, information and data models, in and of themselves, are insufficient to enable the mediation layer to solve the problem of data harmonization in an automated fashion. This is because information and data models are typically defined using informal languages. Hence, a mediation layer that uses just models has no ability to use machine-based reasoning to prove that one set of model elements (e.g., objects, attributes, and relationships) is equivalent to a second set of model elements. For example, in Figure 10.4, there is no way for the mediation layer to *prove* that one set of commands from one router is equivalent to a second set of commands from the second router, even if both data models are derived from the same information model. As another example, refer back

---

[1] Note that the standard relationships, such as associations, aggregations, and compositions that are defined in UML, cannot be used in this example because they lack formal semantics.

to the "Strassner.John vs. JohnS" problem shown in Figure 10.2. Most information models will define a user as a subclass of a more general Person class that has different attributes describing how a Person can be named, such as first-Name and lastName. Given such a set of attributes, how can a data model know whether "Strassner" is a first or a last name? Worse, what does the data model do with "JohnS"? The problem is that there are no *semantics* that can be used to identify what types of attributes these instances are, and no way for the machine to infer that "Strassner.John" is a concatenation of the last name, a period, and the first name of a user.

The goal of FOCALE is to increase the automation of network configuration. Since information and data models by themselves are insufficient to achieve this goal, FOCALE augments their use with ontologies to provide the ability to *learn* and *reason* about management and operational data. In the FOCALE autonomic architecture, knowledge from information and data models is used to represent *facts*. In order to understand the significance and relevance of management and operational data, retrieved facts are converted into a technology-neutral form; knowledge from ontologies is then used to *reason* about facts by defining semantic relationships between model elements and ontological elements.

The Role of Ontologies in FOCALE

Ontologies have their root in philosophy. In philosophy, the study of ontology is a branch of metaphysics that deals with the nature of being. In computer science and knowledge engineering, ontologies define theories of what exists, and they describe the semantics of information. As such, ontologies can be used to define a formal mechanism for representing and sharing knowledge in a machine-understandable form, so that the system can use the knowledge to learn from and reason about sensed data.

There are numerous definitions of ontologies. One of the most popular of these definitions is given in [4], where Gruber defines an ontology as "an explicit specification of a conceptualization." This is expanded on in [5], where a "conceptualization" is defined as the set of objects, concepts, and other entities that are assumed to exist in some area of interest and the relationships that hold among them. When the knowledge of a domain is represented in a declarative formalism, the set of objects, and the set of relationships among them, are reflected in the vocabulary with which a knowledge-based program represents knowledge. In [6], Studer et al. refine Gruber's definition to the following: "An ontology is a formal, explicit specification of a shared conceptualization." In this definition, "formal" refers to the fact that the ontology should be machine readable, and "shared" reflects the notion that an ontology captures consensual knowledge.

In FOCALE, these somewhat abstract concepts were grounded by defining the following goals to be achieved by using an ontology or set of ontologies:

- Each distinct source of knowledge, such as a programming model of a device,[2] is represented by one or more dedicated ontologies; this enables the knowledge present in each knowledge source to be represented using terms specific to that field of knowledge.
- Each ontology must define its set of objects, concepts, and other entities, as well as the relationships that hold among them, using the *same set of declarative formalisms*; this enables applicable parts of each ontology to be combined into a single cohesive whole.
- A set of logical statements that define the terms, rules for relating terms, and the relationships between the terms, of the shared vocabulary of a given ontology.
- A similar set of logical statements that enable multiple ontologies to be related to each other.
- The same formal method for extending the shared vocabulary must be used by all users of each of the ontologies.
- A *life cycle*[3] for ontologies must be rigorously defined.

In summary, FOCALE abides by the principal that "if something is known, it should be able to be represented in a machine-interpretable manner." This gives rise to the following definition of an ontology [7]:

An ontology is a formal, explicit specification of a shared, machine-readable vocabulary and meanings, in the form of various entities and relationships between them, to describe knowledge about the contents of one or more related subject domains throughout the life cycle of its existence. These entities and relationships are used to represent knowledge in the set of related subject domains. Formal refers to the fact that the ontology should be representable in a formal grammar. Explicit means that the entities and relationships used, and the constraints on their use, are precisely and unambiguously defined in a declarative language suitable for knowledge representation. Shared means that all users of an ontology will represent a concept using the same or equivalent set of entities and relationships. Subject domain refers to the content of the universe of discourse being represented by the ontology.

Ontologies are related to cognitive science, where explanations for why certain perceptions, conclusions, and so forth are proposed. Hence, how knowledge is represented is fundamental to the understanding of cognitive issues, such as what a concept means and what implications it has; how it is represented affects any decisions that may arise from that representation. This is of utmost importance in network management, due to the set of heterogeneous data and programming models that abound.

---

[2] Since a device can support multiple programming models and/or multiple languages, the potential exists for an ontology to represent each distinct combination of language and programming model.
[3] A life cycle is defined as the period of time between when an entity is created and when that entity is either retired from service or destroyed.

Knowledge exists in many different forms. Humans, together with the systems and applications that they use, will not "give up" knowledge that they understand or are used to manipulating. Remember the example of an OSS, where the operator has chosen to use multiple systems and components that were not built to interoperate. This lack of interoperability has ensured that different applications and devices develop their own private knowledge bases because there is no easy way to share their knowledge with other applications. This is a serious problem that requires a much more innovative solution than just defining some new tags and representing management data in XML because that will not be able to represent the *semantics* of these heterogeneous knowledge bases. In addition, such an approach would not be able to delineate the set of assumptions made regarding how the application was using that knowledge. The cost of this duplication of effort has been high and will become prohibitive as we build larger and larger systems. Hence, we must find ways of preserving existing knowledge bases while sharing, reusing, and extending them in a standard, interoperable manner.

Information and data models will not and must not go away—they serve a vital need in being able to represent both facts and behavior. Representing knowledge in ontologies enables a formal, extensible set of mechanisms to be used to define mappings between the terms of knowledge bases and their intended meanings. In this way, we can avoid trying to define a single "über-language" that has no underlying business reason, and instead achieve knowledge interoperability by using a set of ontologies to precisely and unambiguously identify syntactic and semantic areas of interoperability between each vendor-specific language and programming model used. The former refers to reusability in parsing data, while the latter refers to mappings between terms, which require content analysis.

Organizing Knowledge Using Ontologies

FOCALE uses the DEN-ng information/data models and the DENON-ng ontologies [8] to represent knowledge. Facts are defined by models; ontologies are used to add semantics to these facts (e.g., provide formal definitions for each fact, as well as a set of linguistic relationships, such as synonyms and meronyms). These data are used to construct finite state machines to represent behavior. Policies are used to determine the set of state transitions that are required when the system is not in its optimal state. Machine reasoning is used to generate hypotheses when a problem occurs; machine learning is used to reinforce actions taken that result in acceptable system behavior.

The design of the DENON-ng ontologies is based on the fact that no single ontology can best answer queries from multiple domains. Therefore, similar to the design of the Cyc ontology [9], DENON-ng is made up of a set of different building blocks that can be combined to form a malleable upper level ontology (i.e., a generic ontology that is used to define common terms that serve as the

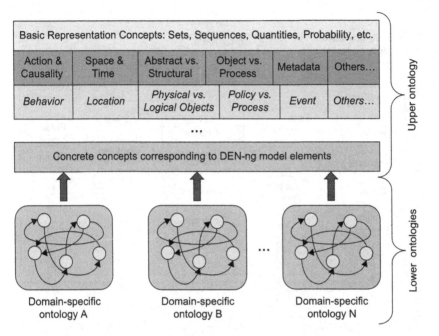

Basic Representation Concepts: Sets, Sequences, Quantities, Probability, etc.

Action & Causality	Space & Time	Abstract vs. Structural	Object vs. Process	Metadata	Others...
Behavior	Location	Physical vs. Logical Objects	Policy vs. Process	Event	Others...

...

Concrete concepts corresponding to DEN-ng model elements

Upper ontology

Lower ontologies

Domain-specific ontology A   Domain-specific ontology B   ...   Domain-specific ontology N

FIGURE 11.1 Conceptual DENON-ng Ontology Design.

basis for integrating multiple lower-level ontologies describing more specific concepts). The most concrete layer of this upper level ontology is a generic ontology that has been designed to augment the main model elements (e.g., classes and associations) of DEN-ng. This inherently extensible layer serves as the root for building domain-specific ontologies, as shown in Figure 11.1. This is another unique aspect of the knowledge representation mechanism used in FOCALE; since it was realized that neither UML-based models nor logic-based ontologies could meet the knowledge representation requirements by themselves, FOCALE embarked on an effort to build a set of ontologies, called DENON-ng, that were constructed as a multilevel set of ontologies designed to use terminology and concepts defined in the DEN-ng information model. This facilitated fusing knowledge from the models with knowledge from DENON-ng using semantic relationships.

DENON-ng is populated, queried, and managed using distributed agents. Each agent is specific to a particular type of knowledge and hence is optimized to process a particular type of knowledge. The ontologies are federated by a set of linguistic and semantic relationships that link each top node of each domain to each other; other lower-level nodes of a domain can of course also be linked to other domain nodes. Figure 11.2 shows an example of the ontology construction process for context data. The ContextData, ManagedEntity, Time, Location, Activity, and PersonOrGroup classes in Figure 11.2 are each linked to similar classes in the DEN-ng model. This enables different domain-specific ontologies

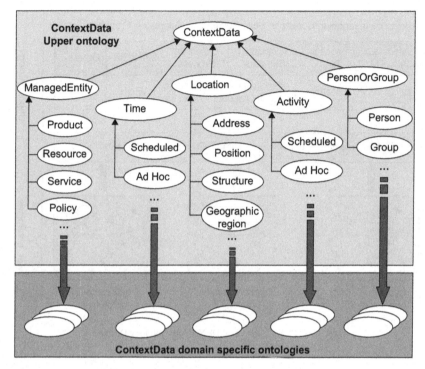

FIGURE 11.2   DENON-ng Ontology for Context.

to be linked to a single higher-level ontology, which in turn enables reasoning over all ontologies to be performed.

Knowledge Integration

Both models and ontologies can be used as references to other types of information. Since information models and ontologies represent data in different ways, graphs are used as a common way to represent their knowledge. The idea is to build a multigraph [10] that conceptually combines information from the models and ontologies into a single larger structure.[4] This is shown in Figure 11.3.

A multigraph is a graph in which more than one edge exists between two nodes. In this approach, the set of graph nodes initially created corresponds to nodes from the models and ontologies. However, there are two types of edges. The first type consists of those defined in the individual models and ontologies, and hence exists between like elements (i.e., model element to model element, or ontology concept to ontology concept). The second type is called

---

[4]Technically, the algorithm also supports other structures, such as hypergraphs and pseudographs, but this is beyond the scope of this chapter.

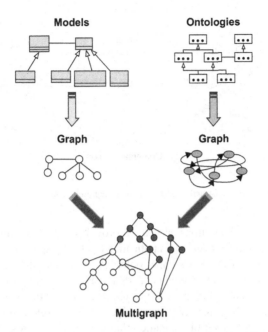

FIGURE 11.3 Constructing a Multigraph to Represent Combined Knowledge.

a semantic edge and exists between different elements (i.e., between model element and ontology element). Semantic edges define the *strength of the semantic relatedness* [11] of the two elements being related to each other.

Semantic relatedness measures how close the meaning of one entity is to the meaning of another entity using one or more lexical relationships, such as synonymy (e.g., "bank" and "lending institution"), antonymy (e.g., "accept" and "reject"), meronymy (e.g., "court" is a part of "government"), and/or other domain-specific relationships. An example of a domain-specific semantic relationship is given in [12]. Hence, the multigraph is used to represent knowledge from each individual model or ontology as well as collective knowledge inferred through the semantic relatedness of elements from the models to elements from the ontologies.

Given that both model data and ontology data are now represented in a consistent way (using a multigraph), semantic relationships can now be defined between each data source by building associations between different nodes and edges in each graph that represent different linguistic relationships, such as synonymy, antonymy, meronymy, or a custom domain-specific relationship. This is done using the notion of a universal lexicon and is shown in Figure 11.4. Step 1 defines a subset of information model nodes N_{IM} from the information model that are the targets for possible semantic enhancement. These targets are identified by mapping concepts embedded in an information model node (or edge, or even a set of nodes and edges) to concepts present in the universal lexicon,

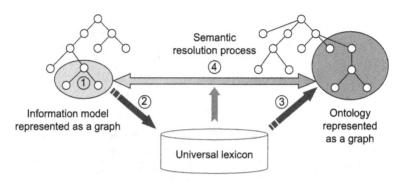

FIGURE 11.4   Constructing a Multigraph to Represent Combined Knowledge.

as shown in step 2. The universal lexicon provides a set of common concepts and terminology that all users and applications of the system have agreed on. It corresponds to the DENON-ng upper ontology. Hence, at this stage, the set of concepts that correspond to model data are translated to a common form so that all applications can use those data. Step 3 uses this information to select a subset of ontology nodes, N_O, from each applicable ontology. Note that in general, an ontology could have a set of semantic relationships that correspond to different meanings of a concept (e.g., "reset" could mean "reboot" or "change to the original value"). In addition, two different ontologies could have two different mappings to the same concept (e.g., a high-level command could have different mappings to different sets of commands for two different vendor-specific devices, each represented by its own ontology). Step 4 defines a set of semantic relationships that relate each unique concept in the set of ontology nodes N_O to each unique concept in the set of model nodes N_{IM}. This is realized as a set of edges $E_{O \rightarrow IM}$, which represents a set of edges that relate knowledge in the ontology to knowledge in the information model. Each edge can be thought of as a pointer that augments the fact modeled in the information model with additional knowledge that exists in the set of ontology nodes that were identified by those edges. The augmentation is of a semantic nature, and serves to clarify and specify the meaning of the concept(s) defined in the model by attaching appropriate definitions and/or linguistic relationships to those model concepts.

During the above four steps, additional relationships and concepts will be discovered. For example, a set of edges $E_{IM \rightarrow O}$ can be defined that relate knowledge in a model to knowledge in an ontology; in this case, the information model would augment concepts in the ontology with facts. This process iterates until either no more relationships can be found or a sufficient number of relationships can be found (this latter case is complex and is beyond the scope of this chapter). Note that, in this figure, ontologies are modeled using an ontology language and are not modeled using UML MOF. Similarly, information and data model knowledge are modeled in UML MOF and are not modeled in an ontology language.

The above process is complicated by the fact that the different types of model elements of the information model each contain different types of data and hence are related to the universal lexicon in different ways. This means that each model element in the information model is mapped to a corresponding model element in an ontology using a combination of linguistic and semantic functions that maximize the semantic relatedness between the information model concept and the ontology concept using concepts defined in the common vocabulary as a reference. Hence, there can be multiple mappings between a given model element and one or more concepts in one or more ontologies. Each mapping represents a different meaning. Therefore, a semantic analysis is done to establish the particular meaning that should be used for the specific context that this mapping is a part of. For complex scenarios, it has been found through experimentation that a set of meanings must be assigned to a particular model element; this means that a set of (possibly different) semantic mappings must be performed for each meaning for each model element. The DEN-ng model, with its novel notion of modeling facts as well as inferences, enables model-driven approaches to be used to help choose which meaning is most appropriate for a given context. In addition, the definition of metadata, as well as differentiating managed entities from simple values, also helps in this selection.

The semantic relatedness for each candidate mapping for a given model element to concept mapping (or vice versa) is then computed and discarded if it does not meet an optional minimum threshold. For all mappings that are kept, the chosen meaning (or set of meanings) from the mapping mechanism is used as an index into each ontology graph to find as many concepts as possible that are related to this meaning. If a term has multiple valid meanings, one semantic relationship is defined for each distinct meaning. In the architecture, this process can get refined and enhanced by using a machine learning algorithm to monitor this process. This enables a machine learning algorithm to specify the degree of semantic similarity (i.e., how close to an absolute match) the semantic algorithm must find in order to produce a useful result; it can also be used to specify how the semantic similarity is computed.

Once all of the algorithm outputs exceeding a programmable threshold equivalence value are determined, a new graph is built. This new graph may be a hypergraph, multigraph, or pseudograph, depending on the nature of the concepts found that map to the output of the modeled data. The new graph contains new nodes and edges that represent new model elements and/or concepts from the ontologies that have been found to be relevant. Note that both positive and negative relationships (e.g., synonyms and antonyms) can be used; many times, it is useful to define what a new concept is not similar to in order to better define what that concept it actually is.

The semantic resolution process compares the meaning (i.e., not just the definition, but also the structural relationships, attributes, etc.) of each element in the first subgraph with all elements in the second subgraph, trying to find the closest language element or elements that match the semantics of the element(s)

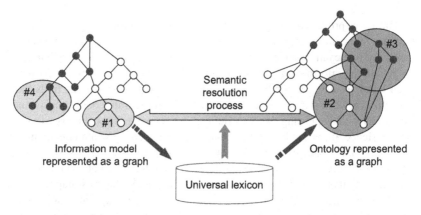

FIGURE 11.5   Constructing a Multigraph to Represent Combined Knowledge.

in the first subgraph. This is an advanced form of reasoning; while other simpler alternatives exist (see [13] for an exemplar list), the quality of the semantic similarity measured will be affected. Often, an exact match is not possible. Hence, the semantic resolution process provides a result in a normalized form (e.g., from 0 to 1), enabling each match to be ranked in order of best approximating the collective meaning of the first subgraph.

These new semantic associations, along with the new elements discovered in the second knowledge source, are used to find new elements in the first knowledge source. The algorithm continues to iterate until either a sufficient number of, or all, significant meanings have been extracted. The overall data structure, then, consists of a set of subgraphs of model elements that are connected to each other that collectively form a new graph, as shown in Figure 11.5.

For example, referring to Figure 11.5, the original subgraph of three nodes and two edges in the information model (shown in ellipse #1), five nodes and three edges (shown in ellipse #2) are added (as a result of the first iteration of matching ontology concepts to information model concepts). Note that only those edges that completely contained in a given domain that is being analyzed (in this case, ellipse #1 and ellipse #2), are counted. Then, another six nodes and six edges (shown in ellipse #3) are added because in this example, the algorithms found new concepts in the ontology that were related to the concepts found previously (in ellipse #2). Finally, four nodes and three edges (shown in ellipse #4) are added by using the algorithm to reverse the mapping (i.e., augment concepts in the ontology with facts from the information model). The end data structure will contain information from all of these nodes and edges. Thus,

Knowledge-Based Governance

Since networks are inherently heterogeneous, their command languages are also different. Even if SNMP is used, in general devices will use proprietary MIBs that either have unique data or represent common data in different ways. Thus,

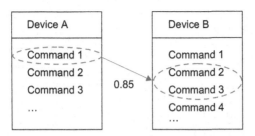

FIGURE 11.6　Mapping Commands between Devices.

an important network management problem is to ensure that the same high-level function executes with the same semantics across multiple devices. This requires that function to be mapped to a set of commands in each device, even though the command structure of each device is, in general, very different.

Ontologies can help solve this problem by defining a *shared* vocabulary, along with rules for defining the structure of the common language. This approach focuses on the structure and formation of words in the vocabulary (morphology), sentence structure (syntax), and meaning and meaning change (semantics).

Figure 11.6 shows an example illustrating this problem. We have a set of M total devices, each having a different number of commands (*i* commands for Device A, and *j* commands for Device M). So, our first objective is to build relationships between one command in one device to a set of commands in another device.

In this example, command 1 of Device A (denoted as A.1) is semantically related to the set of commands 2 and 3 of Device B (denoted as {B.2, B.3} with 85% equivalence (denoted as the "0.85" annotation under the arrow). Note that this is a *unidirectional* mapping. Once a set of these relationships are defined, we can build another mapping that abstracts this to a set of common functions. For example, enabling BGP peering, as shown in Figure 11.4, proceeds as first identifying the high-level function (BGP) to be performed, and then mapping this function to the appropriate set of commands in each device. This is shown in Figure 11.7.

In Figure 11.7, we see that the same high-level function (BGP peering) is mapped to two different sets of commands. This reflects the fundamental fact that different devices have different programming structures and perform the same high-level task using different commands. Significantly, this means that sometimes, the semantics of performing these tasks are different and must be taken into account. Hence, we need an ontology merging tool (or set of tools) to find these semantic differences.

The next task is to determine whether it is *safe* to execute these commands. Often, many different sets of device commands could be used to implement a given high-level function. In general, each set of device commands will have a set of side effects. Hence, the system must be able to evaluate the impact of

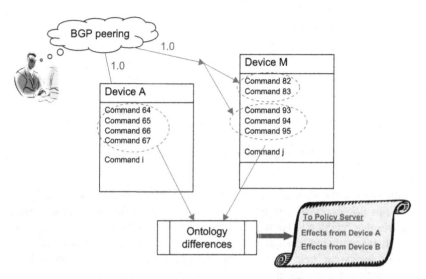

FIGURE 11.7   Mapping a Function to Different Command Sets in Different Devices.

these side effects, so as to choose the best command set to implement. One way to do this is to feed the output of the ontology difference process to a policy-based management system [14]. In this approach, the output of the ontology examination process shown in Figure 11.7 is sent for further analysis by a policy server. The function of the policy server is to define how to implement the device configuration in an optimal fashion.

Figure 11.8 shows a portion of the FOCALE control loops that implement this process. In this figure, a managed resource (which could be as simple as a device interface or as complex as an entire network, or anything in between) is under *autonomic control*. This control orchestrates the behavior of the managed resource using information models, ontology tools, and policy management. Information models are used to map vendor-specific data from the managed resource to a vendor-neutral format that facilitates their comparison. This is accomplished using ontologies, which define a *common vocabulary* for the system. In general, the objective is to see if the actual state of the managed resources matches its desired state. If it does, the top maintenance loop is taken; if it doesn't, the bottom reconfiguration loop is taken. Significantly, ontologies play another role—they enable data from different sources to be compared using its common vocabulary and attached semantic meaning. For example, SNMP and CLI data can be used to ascertain the actual state of the managed resource, even though these are different languages having different descriptions of data. Policies are used to control the state transitions of a managed entity, and processes are used to implement the goal of the policy.

In summary, just as object-oriented class hierarchies form a taxonomy of objects that can be programmed to provide a *function*, information and data

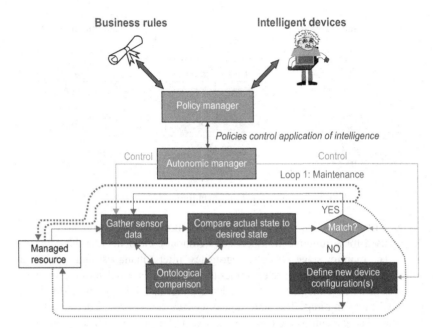

FIGURE 11.8   Policy-based Network Management Using Ontologies.

models form a taxonomy of objects that define *facts* and ontologies form a taxonomy of objects that define the *meaning* of these facts. This combination enables facts gathered by the system (as represented by information and data models) to be understood (using ontologies) and acted upon (using code).

THE EVOLUTION OF THE FOCALE CONTROL LOOPS

The original version of FOCALE [15] was one of the first significant departures from the control loop design pioneered by IBM [16]. This section explains the reasons for this departure and the subsequent evolution of the FOCALE control loops.

The IBM MAPE Control Loop

Figure 11.9 shows IBM's Monitor–Analyze–Plan–Execute, or MAPE, control loop [16]. In this approach, the autonomic element provides a common management façade to be used to control the resource by using a standardized set of interfaces. Its purpose is to define a common management interface that can control a managed entity, regardless of whether that managed entity has autonomic capabilities or not.

The autonomic manager implements the control loop. Sensors and effectors get data from and provide commands to both the entity being managed (bottom portion of Figure 11.9) and to other autonomic managers (top portion of

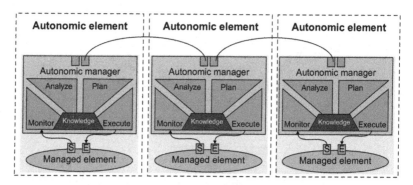

FIGURE 11.9 Conceptual Diagram of IBM's MAPE Control Loop.

Figure 11.9). The set of sensors and effectors forms the management interface that is used for communication between autonomic elements and the environment. The monitor stage collects, aggregates, filters, manages, and reports on sensor data. The analyze stage examines the collected data to determine whether its goals are being met, as well as to learn about its environment and help to predict future situations. The plan stage organizes the actions needed to achieve its goals. The execute stage controls the execution of a plan. The knowledge source implements a repository that provides access to knowledge according to the interfaces of the autonomic manager. In addition, the four control loop functions consume and generate knowledge.

IBM's autonomic computing architecture does not depend on using information or data models or even ontologies, though it can use the CIM in a limited fashion to recognize some types of sensor data. Consequently, the IBM approach is limited to a predefined model of known functionality. In contrast, FOCALE has two models—one of a FOCALE node and one of the environment in which a node is operating. This enables more precise policies and constraints to be developed for each; in particular, it facilitates representing changed functionality when the context of a node changes.

There are several problems with this architecture. First, the control loop is gated by the monitoring function. Hence, if too much data floods the system, the performance of the rest of the system suffers, even if the monitored data is not relevant. Second, the autonomic manager interacts with the entity that it is managing, and not with the environment. Hence, it is relatively difficult for changes in the environment to influence the operation of the control loop. Third, there is no guarantee that an autonomic manager has the ability to perform a desired function, as communication in the IBM architecture depends on the concept of an interface. It is difficult to capture semantics, such as dependencies on other components, side effects, and pre- and postconditions required for a function to be successfully invoked, in the definition of an interface. In contrast, FOCALE uses software contracts [17, 18] to provide interoperable specifications of the functional, operational, and management aspects of a feature.

Fourth, the previous point implies that in the IBM architecture, interfaces represent an exclusive control mechanism and are not intended to be combined with other approaches. In contrast, FOCALE's contracts offer formal mechanisms to combine different information relating to a function. Fifth, it is unclear how distribution is supported in the IBM architecture. Most distributed systems support a set of core services, such as naming, messaging, and transaction management, upon which higher-level services, such as policy management and security, are built; no such services are discussed in the MAPE architecture. In contrast, FOCALE is built as a distributed system and has been the basis of several distributed architectures for various European Framework seven programs. Finally, the IBM architecture uses the autonomic manager as the component to be distributed. As can be seen, the autonomic manager is a relatively complex software entity that is designed to realize a given set of predefined functions. This makes it very difficult for the autonomic manager to support emergent behavior. It also makes it difficult for different autonomic managers to share and reuse knowledge without a common knowledge representation and set of common definitions.

However, the IBM approach suffers from a more fundamental problem: Its control loop is limited to performing *sequential* operations. In this approach, there is no opportunity for data being monitored to be properly oriented toward changing goals. Similarly, the system has no ability to accommodate new or changed information.

The Original FOCALE Control Loops

The FOCALE control loops were originally based on the Observe–Orient–Decide–Act (OODA) [19] control loops developed by Colonel John Boyd. OODA was originally conceived for military strategy but has had multiple commercial applications as well. It describes the process of decision making as a recurring cycle of four phases: observe, orient, decide, and act, as shown in Figure 11.10.

OODA is not a single sequential control loop, but rather a set of *interacting* control loops, where observations in the current context are filtered (the orient phase) to make them relevant. In this approach, orientation shapes observation, shapes decision, shapes action, and in turn is shaped by the feedback and other phenomena coming into the sensing or observing window. The need to orient observations is the inspiration for the FOCALE model-based translation layer, which orients observed data to the current context. *This is one of the primary mechanisms that enable FOCALE to adapt to changes in user needs, business goals, and/or environmental conditions.*

The importance of not using a simple sequential control loop cannot be overstated. First, it is not a good idea to stop observing while the analysis is continuing. Second, a balance must be maintained between delaying decisions (which means delaying actions) and performing more accurate analysis that

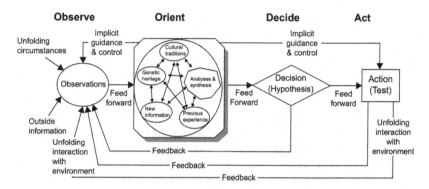

FIGURE 11.10   The OODA Control Loops.

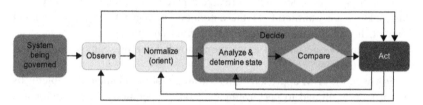

FIGURE 11.11   FOCALE Redrawn as a Modified Set of OODA Control Loops.

eliminates the need to revisit previously made decisions. Third, both the speed of reorientation and the ability to apply suitable actions via the implicit guidance and control link to Action are critical for supporting decision making. This enables a simpler observe–orient–act control loop to be employed in situations that can benefit from this.

The original FOCALE control loops are loosely based on the OODA loops, as shown in Figure 11.11. The Model-Based Translation function corresponds to the Orient function of OODA: It transforms raw sensor data into a form that can be correlated with the current context. The Analyze, Determine State, and Compare functions of FOCALE correspond to the Decide function of OODA, except that OODA is not focused on state, whereas FOCALE is; this is because FOCALE uses state to orchestrate behavior. This is reflected in the Foundation function of FOCALE as well.

The New FOCALE Control Loops

The underlying cognition model of a system is very important, but often is not sufficiently formalized to be fully utilized. While computational performance is increasing, productivity and effectiveness are in fact decreasing for many reasons. One of the most important reasons is that demands on expert users are continually increasing. This has resulted in brittle and rigid systems. An important implication for control loops is that unless they are adaptive, the system that they govern cannot respond effectively to change.

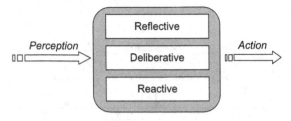

FIGURE 11.12   Reactive, Deliberative, and Reflective Thought Processes.

A cognitive system is a system that can reason about what actions to take, even if a situation that it encounters has not been anticipated. It can learn from its experience to improve its performance. It can also examine its own capabilities and prioritize the use of its services and resources, and if necessary, explain what it did and accept external commands to perform necessary actions. Minsky modeled this system using three interacting layers, called reactive (or subconscious), deliberative, and reflective [20]. These processes are shown in Figure 11.12.

The FOCALE control loops were redesigned to incorporate advances in cognitive psychology in order to have the autonomic manager more closely model how humans think. In this approach, reactive processes take immediate responses based on the reception of an appropriate external stimulus. It reacts in order to carry out one or more goals. If there is a difference between the received data and carrying out the goal, it shifts the attention of the person accordingly. It has no sense for what external events "mean"; rather, it simply responds with some combination of instinctual and learned reactions. Deliberative processes receive data from and can send "commands" to the reactive processes; however, they do not interact directly with the external world. This part of the brain is responsible for our ability to achieve more complex goals by applying short- and long-term memory in order to create and carry out more elaborate plans. This knowledge is accumulated and generalized from personal experience and what we learn from others. Reflective processes supervise the interaction between the deliberative and reactive processes. These processes enable the brain to reformulate and reframe its interpretation of the situation in a way that may lead to more creative and effective strategies. It considers what predictions turned out wrong, along with what obstacles and constraints were encountered, in order to prevent suboptimal performance from occurring again. It also includes self-reflection, which is the analysis of its own performance, as well as how well the actions that were taken solved the problem at hand. The new cognition model makes a fundamental change to the FOCALE control loops, as shown in Figure 11.3 and Figure 11.14.

Instead of using two alternative control loops (one for maintenance and one for reconfiguration), of which were reflective in nature, the new FOCALE architecture uses a set of outer control loops that affect the computations performed in a corresponding set of inner control loops. The outer control loops are used

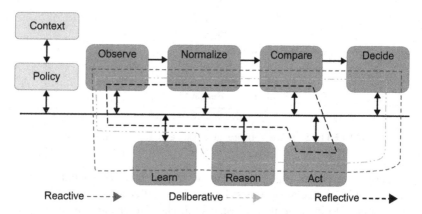

FIGURE 11.13   The New FOCALE Control Loops: Inner Loops View.

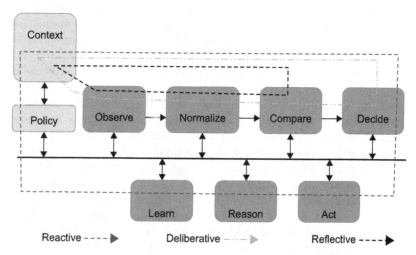

FIGURE 11.14   The New FOCALE Control Loops: Outer Loops View.

for "large-scale" adjustment of functionality by reacting to context changes. The inner control loops are used for more granular adjustment of functionality within a particular context. In addition, both the outer and the inner control loops use reactive, deliberative, and reflective reasoning, as appropriate.

The inner loops have two important changes. First, the Decide function has been unbundled from the Act function. This enables additional machine-based learning and reasoning processes to participate in determining which actions should be taken. Second, the new cognition model has changed the old FOCALE control loops, which were reflective in nature, to the three control loops shown in Figure 11.13, which are reactive, deliberative, and reflective. This streamlines analysis and processing, freeing the resources of an individual FOCALE autonomic element to be used for other purposes, as discussed in the next section.

Figure 11.14 shows the new outer loops of FOCALE. The reactive loop is used to react to context changes that have previously been analyzed; the deliberative loop is used when a context change is known to have occurred, but its details are not sufficiently well understood to take action; and the reflective loop is used to better understand how context changes are affecting the goals of the autonomic element.

SUMMARY

This chapter has described the design and evolution of the FOCALE autonomic architecture. FOCALE has been designed for managing heterogeneous networks. However, its principles can be used for other applications, as the mechanisms described are not specifically oriented toward network management. In particular, the most difficult challenge that FOCALE has overcome—integrating heterogeneous management and operational data from different sources that use different languages to describe their data—makes FOCALE well suited to being used in other industries, since the integration of disparate management and operational data was done using a novel new knowledge representation that was independent of any underlying technology being managed.

This chapter has emphasized the importance of knowledge representation and fusion. Too often, network management focuses on a specific technology or device, or a set of technologies and devices, and ignores the larger picture of how network services and resources are used by different users, as well as how environmental conditions affect the network services and resources offered. The future will increasingly be dominated by the need to enable services that have conflicting requirements of shared resources to coexist. In addition, context awareness—the ability for the resources and services offered at any given time to *change* as a function of user needs, environmental conditions, business goals, or other factors—must be taken into account. Static governance approaches that use static knowledge bases cannot deal with these complexities.

The FOCALE approach recommends integrating models and ontologies. Models provide a powerful way to describe facts and support advanced code generation methodologies. Ontologies represent a powerful set of tools to attach semantic meaning to data. The FOCALE architecture is a specific architecture that uses models and ontologies to represent knowledge. The Model-Based Translation module converts vendor-specific sensor data into a common vendor-neutral form, and vendor-independent commands into vendor-specific versions. The autonomic manager determines the state of the entity being managed from received sensor data and/or inferences, and determines what actions to take guided by context-aware policy rules. Semantic reasoning helps orchestrate system behavior through the use of machine-based learning (which enables system operation to be improved with experience) and reasoning (which generates hypotheses of root causes of problems and proves whether or not a hypothesis is correct).

REFERENCES

[1] J. Strassner, "Enabling autonomic network management fecisions using a novel semantic representation and reasoning approach," Ph.D. thesis, 2008.

[2] J. Fuentes, V. Quintana, J. Llorens, G. Genova, and R. Prieto-Diaz, "Errors in the UML metamodel," ACM SIGSOFT Software Engineering Notes, Vol. 28, No. 6, November 2003.

[3] R. Neches, R. Fikes, T. Finin, T. Gruber, R. Patil, T. Senator, and W. Swartout, "Enabling technology for knowledge sharing," AI Magazine, Vol. 12, No. 3, pp. 36–56, 1991.

[4] T.R. Gruber, "A translation approach to portable ontology specification," Knowledge Acquisition, Vol. 5, No. 2, pp. 199–220, 1993.

[5] http://www-ksl.stanford.edu/kst/what-is-an-ontology.html.

[6] R. Studer, V. Benjamins, and D. Fensel, "Knowledge engineering: Principles and methods," IEEE Transactions on Data and Knowledge Engineering, Vol. 25, No. 1–2, pp. 161–197, 1998.

[7] J. Strassner, "Knowledge engineering using ontologies," Handbook of Network and System Administration, edited by J. Bergstra and M. Burgess, Chapter 3, pp. 425–457, 2007.

[8] J. Serrano, J. Serrat, J. Strassner, G. Cox, R. Carroll, and M. Ó Foghlú, "Policy-based context integration & ontologies in autonomic applications to facilitate the information interoperability in NGN," 2nd Workshop on Hot Topics in Autonomic Computing, pp. 6–16, June 15, 2007, Jacksonville, FL.

[9] http://www.cyc.com.

[10] R. Diestel, Graph Theory, 3rd ed., Springer Graduate Texts in Mathematics Series, 2006.

[11] E. Gabrilovich and S. Markovich, "Computing semantic relatedness using Wikipedia-based explicit semantic analysis," Proc. of the 20th Intl. Joint Conference on Artificial Intelligence, pp. 1606–1611, 2007.

[12] A. Wong, P. Ray, N. Parameswaran, and J. Strassner, "Ontology mapping for the interoperability problem in network management," IEEE Journal on Selected Areas in Communications, Vol. 23, No. 10, pp. 2058–2068, October 2005.

[13] P. Shvaiko and J. Euzenat, A Survey of Schema-based Matching Approaches, Technical Report DIT-04-087, Informatica e Telecomunicazioni, University of Trento, 2004.

[14] J. Strassner, Policy-Based Network Management, Morgan Kaufman Publishers, September 2003.

[15] J. Strassner, N. Agoulmine, and E. Lehtihet, "FOCALE—A novel autonomic networking architecture," International Transactions on Systems, Science, and Applications (ITSSA) Journal, Vol. 3, No. 1, pp. 64–79, May 2007.

[16] IBM, "An architectural blueprint for autonomic computing," June 2006, 4th ed., available for download from: http://www-01.ibm.com/software/tivoli/autonomic/pdfs/AC_Blueprint_White_Paper_4th.pdf.

[17] R. Mitchell and J. McKim, Design by Contract, by Example, Addison-Wesley, 2001.

[18] S. van der Meer, J. Strassner, and P. Phelan, "Architectural artefacts for autonomic distributed systems—Contract language," 6th IEEE Workshop on Engineering of Autonomic and Autonomous Systems (EASe), April 14–16, 2009.

[19] J. Boyd, "The essence of winning and losing," June 28, 1995, available for download from: http://www.projectwhitehorse.com/pdfs/boyd/ The%20Essence%20of%20 Winning%20and%20Losing.pdf.

[20] M. Minsky, The Society of Mind, Simon and Schuster, New York, 1988.

Conclusion

The concepts of autonomic network management and autonomic networks have been proposed to address the problem of increasing complexity in existing networks as well as the corresponding increase in cost. Autonomic network management and autonomic networking aim at freeing human networks and systems administrators from the complexity of managing existing and future network infrastructure and services. The promising idea of developing new generations of physical and logical network elements that are able to manage and organize themselves to address expected and unexpected situations is part of the vision of all administrators and network providers. In this approach intelligence is shifted from the human brain directly into the autonomic elements, allowing the network to work in a closed control loop (i.e., to monitor, analyze, plan, and execute) to achieve its objectives.

Very little has been achieved so far toward realizing this ambitious vision of not only future networks but also the whole information system. Existing works have addressed just some aspects of these challenges, solving only the visible part of the iceberg. There is indeed a need to rethink the whole way network components and services are designed and to tailor them so that they're inherently self-managed to build future networks. This issue is addressed by several initiatives under the umbrella of Future Internet or Next Generation Networks in the United States, Europe, or Asia. New types of software and hardware components need to be designed in such a way that they can seamlessly communicate, integrate, and organize solely into a complete system that is able to manage itself with little or no human intervention. Network protocols and information models should be able to evolve and adapt in a seamless manner to changing contexts. With the emergence of new types of applications and communication technologies, such as ad-hoc networking, wireless sensors networking, and cognitive networks, network components should have even higher adaptation capabilities and be able to cope with various contexts.

The various contributions presented in this book aim at presenting different theoretical and practical directions toward building such self-managed networks. Some initiatives have focused on the network management aspects, while others have concentrated on network communication. The common denominator among all these approaches is the introduction of self-* capabilities in the network, implementing a closed control loop as a means to achieving self-governance, freeing human administrators from mundane tasks, and allowing them to focus on business processes management.

Autonomic Network Management Principles. DOI: 10.1016/B978-0-12-382190-4-00012-7

The journey to a fully autonomic network is an evolution comparable to autonomic computing and its five stages presented by IBM (see Chapter 1). Moreover, we envision that initiatives focusing on autonomic network management and autonomic networks will converge in the long run but focus will remain slightly different in the short and middle term as they could go through different intermediate stages as depicted in the following figure:

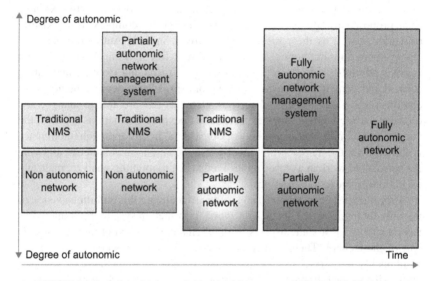

FIGURE 1   Evolution of Autonomic Network Management and Autonomic Networks.

Autonomic network management solutions will be characterized by different aspects that highlight their degree of autonomic maturity:

1. The degree of autonomy within a given scope, that is, how self-management properties are introduced in the network and services components. This is very important as it will determine the operation cost that the operator could save either by reducing the costly human intervention or simplifying the life cycle of the newly introduced network components and services. The operator will achieve a faster return on investments and will ensure a higher service quality to its customers (availability, performance, security, etc.).

2. The degree of adaptability, that is, the capacity of the solution to adapt to new situations either with prior knowledge of the situation or not. This degree of adaptability can be achieved in the first stage using new modeling design, but the main evolution could be achieved by following emerging biologically inspired technologies. Bio-inspired techniques are gaining more and more interest during recent years, and a better understanding of these techniques will certainly benefit the networking area and will allow the development of autonomic networks that have the same intrinsic adaptation capabilities as biological systems do. Nevertheless, these approaches will require that

scientists from different areas work together toward the creation of new system design approaches that will be used in the area of network design and management.

3. The degree of learning and diagnosis. Adaptability can be efficient only if the autonomic network is capable of correctly interpreting the information it gathers about its environment and about its own behavior. Autonomic network management systems need to evolve during their lifetime by learning about situations and the way problems are solved without always referring to predetermined problem-solving schemas. Learning is a crucial property needed to achieve autonomic network management. By enhancing their knowledge, autonomic network management systems can address new situations in a more efficient way, always finding the best configuration to improve its performance, security, and the like.

4. The degree of abstraction. Existing programming languages have been created to develop systems that were not thought to be fully autonomic. To facilitate the development of autonomic networks and corresponding self-management capabilities, new abstractions need to be created to address the challenge of autonomic behavior and adaptation. The range of new programming approaches will require innovation in the core of programming language design and not simply new libraries using classical programming style.

The following figure presents the expected generations of ANMS and AN based on the maturity of the concepts that have been used to build these systems, namely, degree of autonomy, degree of learning and diagnosis, degree of adaptability, and degree of abstraction.

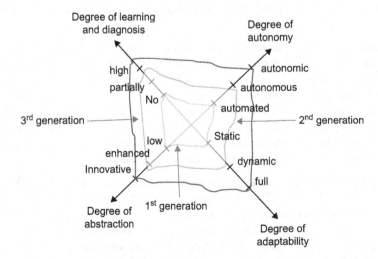

FIGURE 2   Different Dimensions of Autonomic Network Management Systems.

The contributions presented in this book address these different dimensions and propose different approaches to autonomic network management and autonomic networks. Obviously, the proposed solutions are not the ultimate ones but they will pave the way to future fully autonomic networks.

The following summarizes the contributions of each chapter.

Chapter 1, "Introduction to Autonomic Concepts Applied to Future Self-Managed Networks." This chapter presents a general overview of autonomics applied to networking and to network management. Future autonomic network management systems and autonomic networks must use innovative approaches that do not always necessarily represent new technologies but certainly more convergence between different domains (computer science, networking, bio, economy, sociology, cognitive, etc.). It is through the convergence of ideas from different domains that innovative solutions will be created. Autonomy requires more intelligence, more learning, more adaptation, more organization, and more awareness to be really efficient in contexts that are more and more complex such as networks and services.

Chapter 2, "Autonomic Overlay Network Architecture." The proposed management scheme presented in this chapter provides an integrated architecture for autonomic SSONs management. The envisioned layered architecture for autonomic overlay provision enables autonomy and dynamic overlay construction through multilevel policies. The idea of components that are able to self-assemble into an overall autonomic system frees the administrator from any manual intervention. SSONs have the ability to customize the virtual network topology and the addressing, as well as the routing at the overlay level, according to the specific requirements of a media delivery service. As the authors conclude, the road toward fully autonomic system architecture is still long; however, this chapter presents an interesting approach for the design of future autonomic services.

Chapter 3, "ANA: Autonomic Network Architecture." This chapter describes the reference model of the ANA and the underlying basic abstractions and communication paradigms. The results presented in this chapter have shown the benefit of using the Compartment as the basic unit to design autonomic network and a means to support interaction and federation. The authors have shown that the proposed design using Compartments helps to model autonomic networks more easily and to ensure manageability, as each compartment is in essence full autonomy capable of handling its internal communications.

Chapter 4, "A Utility-Based Autonomic Architecture to Support QoE Quantification in IP Networks." This chapter presents a solution to build an autonomic network management system with self-decision capabilities. The approach followed by the author using the Utility Theory helps to define optimization functions that allow the ANMS to make decisions on its own based on a collection of monitored data. The authors conclude that the approach sounds promising, but the mechanisms to transform the high-level human requirements into objectives expressed in terms of Network Utility Functions is still an open issue and requires more investigation in the context of a real network.

Chapter 5, "Federating Autonomic Network Management Systems for Flexible Control of End-to-End Communication Services." As in traditional network management, end-to-end network management requires NMSs from different administrative domains to interact in order to ensure end-to-end services. This is also the case in autonomic network management systems. Nowadays network management systems do federate to provide end-to-end delivery of communications services. The proposed solution highlights the benefit of a Layered Federation Model that attempts to encapsulate and interrelate the various models, processes, and techniques that will be required to realize such a flexible approach to management system federation. This work will certainly open up new challenges in the future, as heterogeneity at the administrative level and interoperability between network operators to provide autonomic end-to-end services is an important issue in the real world.

Chapter 6, "A Self-Organizing Architecture for Scalable, Adaptive, and Robust Networking." This chapter proposes a self-organizing network architecture where each node, network, layer, and network system is self-organized through intra- and interlayer mutual interactions inspired by biological systems. The approach shows that self-organization and self-adaptation emerge while following these principles. However, fully decentralized decision making does not necessarily guarantee the best performance. It could be worthy to sacrifice performance to some extent to achieve scalability, adaptability, and robustness in large-scale networks.

Chapter 7, "Autonomics in Radio Access Networks." Autonomic management in the area of radio access networks is one of the most important challenges for wireless networks operators. The contribution presented in this chapter addresses the general trend of next-generation RANs, which is to provide better and more attractive services, with growing bit rates to support high-quality multimedia applications. Autonomic principles have been applied in this work to provide an efficient autonomic solution for dynamic resource allocation in the RAN providing higher spectral efficiencies at lower operation costs. In this context, radio systems with SON capabilities. The proposed solution shows that learning is a central mechanism in autonomic RAN management systems. The proposed statistical learning approach along with RL has provided promising results to achieve self-diagnosis, self-healing, and self-optimization tasks. This solution could really inspire other approaches in similar areas.

Chapter 8, "Chronus: A Spatiotemporal Macroprogramming Language for Autonomic Wireless Sensor Networks." As previously stated, programming paradigms are important for the design of future autonomic networks and network management. The programming paradigm for autonomic WSNs proposed by the authors is an important contribution in the area, for it helps to reduce the complexity of programming autonomic WSN evolving in two-dimensional space. Indeed, the Chronus language has been designed to address two-dimensional physical space sensor nodes. It is planned to be extended to three-dimensional (i.e., four-dimensional spacetime). Initiatives in this direction will allow development of autonomic WSN addressing three-dimensional space and providing application such as building monitoring and atmospheric monitoring.

Chapter 10, "The Design of the FOCALE Autonomic Networking Architecture." While IBM's autonomic computing architecture provides a framework for designing autonomic computers, it is not sufficient for the design of autonomic network management systems. Indeed, network management requires additional concepts to address the specificities of network physical and logical elements' management. The FOCALE architecture is a specific architecture that uses models and ontologies to represent knowledge. The Model-Based Translation module is at the heart of the architecture, as heterogeneity is present at each level of the network stack and needs to be solved. The proposed FOCALE architecture provides additional modules to address these challenges and constitutes a reference architecture for future ANMS.

Chapter 11, "Knowledge Processing in FOCALE." Integrating heterogeneous management and operational data from different sources that use different languages to describe their data makes the proposed FOCALE architecture very attractive. Indeed, disparate management and operational data are integrated using a new knowledge representation that is independent from any underlying technology being managed. In addition, context awareness—the ability of resources and services offered at any given time to change according to user needs, environmental conditions, business goals, or other factors—must be taken into account. The FOCALE approach recommends integrating models and ontologies. Models provide a powerful way of describing facts and support advanced code generation methodologies. Ontologies represent a powerful set of tools to attach semantic meaning to data.

To create truly autonomic network management systems and autonomic networks, scientists must consider much more radical approaches and shift from traditional paradigms. In this book, we have presented several contributions aiming at introducing autonomic functions in the design of network management systems or directly in the network core. The approaches are not proposing new technologies to radically change system design but rather to combine technologies from different areas of IT, as well as from other areas of science such as biology to build network management systems and sometimes networks with self-management capabilities. All these contributions help us to understand the area better and give the readers some ideas about how future autonomic network management systems and autonomic networks will be built.

Index

Printed in the United States
By Bookmasters